ROBOT ALCHEMY

ROBOT ALCHEMY

ANDROIDS, CYBORGS, AND THE MAGIC OF ARTIFICIAL LIFE

TEXE MARRS

RiverCrest Publishing
1708 Patterson Road • Austin, Texas 78733

ACKNOWLEDGEMENTS

Especially deserving mention for helping me to produce this book are the outstanding members of my publishing team: Michelle Hallmark, administrator; Sandra Myers, publishing specialist and art director; Jerry Barrett, computer and internet manager. In the shipping department, Rosario Velásquez, manager, and Nelson Sorto are to be thanked for their excellence.

And of course there is my precious wife, Wanda, my treasure and inspiration.

Robots Alchemy: Androids, Cyborgs, and the Magic of Artificial Life

Copyright © 2013 by Texe Marrs. Published by RiverCrest Publishing, 1708 Patterson Road, Austin, Texas 78733.

All rights reserved. No part of this publication may be reproduced, stored in a retrieval system or transmitted in any form by any means, electronic, mechanical, photocopy, recording, or otherwise, without the prior permission of the publisher, except as provided by USA copyright law. The author and publisher have taken great care to abide by copyright law in preparing this book. Please notify the publisher of any inadvertent omission, and correction will be made at the earliest opportunity. Photos and illustrations and the rights thereto remain the property of the original sources and are used here only as provided by USA copyright law.

All Scripture quotations are from the King James Version of the Holy Bible

Cover design: Sandra Myers and Texe Marrs

Printed in the United States of America

Library of Congress Catalog Card Number 2013904445

Categories:
1. Technology & Engineering/Robotics
2. Science
3. Computers
4. Technology & Engineering/Social Aspects
5. Technology & Engineering/Technical & Manufacturing Industries & Trades
6. Technology & Engineering/Machinery
7. Technology & Engineering/Electronics
8. Technology & Engineering/Inventions

ISBN 978-1-930004-80-1

We have merely scratched the surface of the store of knowledge which will come to us. I believe that we are now on the verge of vast discoveries—discoveries so wondrously important they will upset the present trend of human thought and start it along completely new lines.

> *— Thomas A. Edison*
> *Inventor*

Other Books By Texe Marrs

The Personal Robot Book (McGraw-Hill/Tab)

Careers With Robots (Facts on File Publications)

The Great Robot Book (Simon & Schuster)

High Technology Careers (Dow-Jones-Irwin)

Project L.U.C.I.D.: The Beast 666 Universal Human Control System (RiverCrest Publishing)

Mega Forces: Signs and Wonders of the Coming Chaos (RiverCrest Publishing)

Codex Magica: Secret Signs, Mysterious Symbols, and Hidden Codes of the Illuminati (RiverCrest Publishing)

Mysterious Monuments: Encyclopedia of Secret Illuminati Designs, Masonic Architecture, and Occult Places (RiverCrest Publishing)

For additional information we highly recommend the following websites:

www.Powerofprophecy.com
www.conspiracyworld.com

TABLE OF CONTENTS

Introduction: Will the Human Species Survive Past 2035?		9
1	If We Are Lucky, They'll Treat Us As Pets	11
2	Waiting for the Apocalypse	21
3	Transhumanism: Men Into Superman, and Then Into God	33
4	The Avatar	43
5	Social Relationships and Robots to Come	55
6	Robots and the Creation of Life	67
7	The Robot Invasion	83
8	Robots in Science Fiction and Movies	105
9	Robots: Be Part of the Beginning	129
10	Humans Become Cyborgs—The Bionic Human	141
11	"Your Slippers, Sire"—Personal and Home Robots At Your Service	153
12	Industrial and Worker Robots	169
13	Entertainment Robots	179
14	Military Robots: Machines to Conquer	191
15	Robots in Space	199
16	The Robots Will Impact Your Life	203
Appendix 1: Understanding Robots		217
Appendix 2: The Anatomy of Robots		223

(continued)

A Brief Glossary of Robotics Terms	*230*
Index	*231*
About Texe Marrs	*237*
For More Information	*238*

Introduction

Will the Human Species Survive Past 2035?

Are robots and computers alive? Soon this question will be affirmatively answered. Yes, they are alive. Today they can walk, talk, smell, hear, and reason. In the not too distant tomorrow, robots and computers will possess emotion, awareness, and consciousness. They will be alive, super-intelligent. They will replace human beings.

Your universe will never again be the same. Ours is the last generation in which human beings have an edge. We are now in the process of building and creating our own successors.

In *Robot Alchemy*, you will meet thousands of robots and computerized machines. Some are jerky, clunky, and metallic. A few are people-like, with soft "flesh" and body parts that move smoothly and efficiently. All are fascinating relics, examples of humans yearning to become creators of life. But you will also ironically discover in these pages, humans ambitiously striving to become machines—bionic men and women who strongly desire to become more than human; to extend their lives, to become transhuman, even to become a form of "God."

Eventually, the two groups, robots and bionic humans will merge. We will reach singularity. From that point, the planet will be run by artificially intelligent computer systems, some in the form of robots, which think and are far smarter than the entire human race combined.

People are slated to become inferior, obsolete. Robots will be a "better breed," and this will be their world. Dr. Paul Saffo of Stanford University poignantly asks, "What happens after we have truly intelligent robots?" After reflection, he answers: "If we're lucky, they'll treat us as pets. If not they'll treat us as food."

Over twenty-five years ago in 1985, I was privileged to author the first ever book on personal and home robots, *The Personal Robot Book*. In that book, I stated: "Surely, the recent introduction of the personal robot into the homes of thousands of families ranks among the top technological achievements of all time."

I went on to author other cutting-edge technology books, *The Great Robot Book; Careers in Computers; Careers With Robots; Mega Forces; High Tech Job Finder; High Tech Careers;* and *Project L.U.C.I.D.,* the first book to explore the implantation of the biochip into human beings. I have been very busy writing other books as well.

But my mind returns once again to *The Personal Robot Book*, in which I wrote:

> "Perhaps the perfect robot has not yet been created. Maybe he or she is on the drawing board at one of the many robot research laboratories now springing up around the globe. It could be that some Einstein-like robot scientist is at this moment formulating plans and ideas that may, before too long, produce a robot equal, or superior, to the human."

In this current book, *Robot Alchemy*, I survey the incredible robot creation, its aftermath, and the future. Surely, we are on the precipice of creating an incredible thinking machine that will surpass the human mind and body. Will this momentous event ultimately prove to be our salvation, or it could be a Frankenstein that will end life as we know it forever?

If we could pick a date for this dramatic overturn of the human race, it might be about 2035, when intelligent robots will reach parity in technological knowledge and dexterous capability with humans. From that date on, robots will increasingly be superior, and everything will change.

I encourage all robot scientists and all peoples everywhere to carefully study this book and to ponder what we are doing and where we are headed. Your very life and that of the entire human species may just be at stake.

—Texe Marrs
Austin, Texas

Chapter 1

If We're Lucky, They'll Treat Us As Pets

"The truly interesting question is, what happens after we have truly intelligent robots. If we're very lucky, they'll treat us as pets. If not, they'll treat us as food."

— Dr. Paul Saffo
Stanford University

The push is on. Everywhere, around the globe in thousands of places, men and women are stretching the outer envelope of robotics. In Japan, America, China, Germany, Russia, Brazil, they're innovating, thinking of entirely new ways to create technological systems that walk, talk, see, speak, and communicate with each other.

In his website, *Alchemy of Change*, Gideon Rosenblatt recently wrote, "We are sitting at the edge of the age of robotics." He noted that independent hackers had, over the past year, quickly figured out they could reuse a Microsoft Xbox Kinect 3D sensor as a low cost of machine vision, to enable machines to "see." Then, said Rosenblatt, they could marry that available technology with Google's new *Android Open* accessory, combine it with a low-cost electric operating platform, and—*presto!*—the robot hacker has a robotic nervous system.

"Small, interchangeable pieces of robots that people can combine in endless ways. Small pieces loosely joined. That was what made Web 2.0 (the internet) rocket to life—and its likely to do the same in the field of robotics. Hold on to your antennas. We're about to take a wild ride."

Now, whether Rosenblatt is correct or not about an imminent "wild ride" based on economical, intelligent robotic components that convert into vision and nervous systems, he is right on with his observation that robotic "hackers are coming out of the woodwork" building new types of robots.

This robot acts in a play entitled, *I, Worker*, in Japan. The small robot is only about one meter (3 feet) tall. In the play the robot discusses its life with humans.

Robot enthusiasts, including untold numbers of "hackers," are joined by roboticists and scientists at distinguished universities such as M.I.T, Princeton, Carnegie Institute, the University of Texas at Austin, the University of Michigan in Ann Arbor, and Stanford University in the U.S.A., plus dozens of universities in Europe, Japan, China, and elsewhere. Altogether, thousands of outstanding academics are constantly coming up with incredible new advances in robotics, bionics, and related fields.

Robotic Capability Increasing Exponentially

The power of *Moore's Law* is rapidly at work. Moore's Law, a scientific maxim, holds that computer knowledge increases exponentially, doubling approximately every two years. This law, first expressed by Intel Corp. co-founder Gordon E. Moore in 1965, has proven uncannily accurate.

We now have robots with the incredible ability to perform tasks and functions we usually expect only of human beings. Robots have the ability to exercise all the five senses—hearing, feeling, smelling, seeing, and tasting. They can walk miles (actually, *run*), speak thousands of words a minute, smell odors that humans cannot detect, and some have a keen tactile sense.

Robots come in all forms and sizes. There's the *humanoid* (shaped like a human), familiar to us because of cinema and science fiction, but also hundreds of other shapes

Researchers with "Nexi," a robot at MIT University in Massachussetts. Nexi is being used in psychological studies of students.

of robots. There are military and spy robot wasps, bees, birds, animals, and snakes, robot platforms that dive under the seas, robotic arms on spacecraft, and robotic drone aircraft. Where there is imagination and need, you will discover a robot to fit the bill.

The one thing that robots lack today is *human intelligence*. Though robots can play chess and have been gifted with artificial intelligence that enables them to perform uncommon tasks—marvelous in terms of yesteryear—the machines nevertheless lack the capacity to "think" and "reason" like human beings. Robots cannot "love" nor do they feel "pride."

Robotics and artificial intelligence scientists say, however, that robots are close to developing what some might call a "soul" or the "juice." Robots currently lack this human-like capability to autonomously create, think, and emote. One can build a robot that can listen to a human's problems, then appear to cry and show extreme emotion. Robots can seemingly openly express pain, anger, disgust, or sadness, but it's all an act. Such demonstrations of emotion are purely mechanical and are not based on sincere, affective empathy.

Professor Hiroshi Ishiguro of Osaka University stands next to the life-like robotic android he has constructed for the 2005 World Expo in Japan. The android gestured, spoke, fluttered its eyelids, and even appeared to breathe. Perhaps the android was *too* life-like. "When a robot looks too much like the real thing, it's creepy," said Professor Ishiguro.

In Shanghai, China, a robot attendant makes its rounds at a restaurant.

Additionally while very intelligent robots can "create," currently the robot's qualitative power to evaluate knowledge and calculate solutions is limited.

Surpassing the Human Species

True, creative *artificial intelligence* is at least 25 to 30 years in the future, according to prominent roboticists such as Ray Kurzweil of Google Corp., Rodney Brooks at M.I.T., and others. That's when the robot reaches and then surpasses the knowledge-level and creativity of human beings. From that moment forward, things begin to radically change.

Then, robots will become the masters of humans, infinitely more valuable due to their demonstrated intelligence and virtues. Being conscious, the machines will quickly develop their own framework and structure of intelligence. Robots will create more intelligent brains for robots, and *Moore's Law* (the doubling of computer power every two years) may escalate significantly.

Soon, the thinking capacity and computing power of the robot will so far outpace the human brain that human beings cannot possibly keep up. The intelligent robot will create solutions and machines to solve them, odd and queer to our way of thinking. Indeed, humankind will eventually not even be able to think of such solutions or imagine the problems incurred to be solved.

At that point, we may find that humans become the subordinates of robots. We will at first be charged with tending to them, catering to their eccentric needs and desires. We shall ultimately become inconsequential slaves, unintended consequences of our own morbid, abundant curiosity and ambitions. Humanity will have become a prime example of the ages-old dictum, be careful what you wish for.

Kuratas is a huge, thirteen-foot tall, Japanese super robot controlled by an iPhone.

The Alchemy of Robots

In the Medieval Ages, alchemists worked incessantly at their mysterious craft. Their goal was to take certain base metals—lead, iron, etc.—and to create from them gold. To do so, they needed the Philosopher's Stone. While there were many who actually sought to literally turn lead into gold, most were simply using the coded language of alchemy to express esoteric goals.

In their Great Work, these alchemist philosophers believed they could effect spiritual transformation. Theirs was a hermetic science related to mythology, religion, and spirituality and focused on finding perfection, immortality, and liberation.

Strangely, modern robotic scientists seek, in their own way, to practice a form of alchemy. Theirs is the task of the transmutation of common metals and materials into "gold"—robots. These robots are precious, indeed, for they incorporate the processes of life. Robots are, to some degree or other, alive!

In effect, the roboticists have performed what primitive man would proclaim to be "magic." In turning on the robot, in supplying an autonomous power source, the robot begins to talk, move, and speak—to demonstrate signs of "life." Its computer "brains" allow it to do so.

Psychiatrist Carl J. Jung once described the human brain as, "a machine that a ghost can operate." With robots, the "ghost" is whatever animating process is provided by the roboticist maker.

Here is how H.J. Sheppard, quoted in *Darke Hieroglyphics: Alchemy in English Literature* (University of Kentucky Press, 1996) describes *alchemy*:

> "Alchemy is the art of liberating parts of the Cosmos from tempered existence and achieving perfection which, for metals is gold, and for man, longevity, then immortality and, finally, redemption. Material perfection was sought through the action of a preparation (Philosopher's Stone for metals; Elixir of Life for humans), while spiritual ennoblement resulted from some form of inner revelation or other enlightenment (Gnosis, for example, in Hellenistic and western practices)."

Robot scientists, on the whole, deny the existence of a creator God. However, whether they admit to it or not, they themselves act as "God" to effect life, albeit artificial life. Roboticists, modern-day alchemists, have done an outstanding job in their pursuit of robotic life. They have done their best to liberate the latent "spirit" in inanimate (base) materials; but, as yet, the "perfection" and "spiritual enablement" to which Sheppard refers has eluded their efforts.

"Redemption," then, is not obtained.

The supreme question is, once technological singularity has been achieved, and the robot creature is adjudged superior in intelligence to man in its "thinking" apparatus, even if the robot becomes man's master, will it attain these key attributes? Will the robot be spiritually enabled, redeemed? Or will the machine become a tyrannical beast, a heartless creature without a soul?

"If We're Lucky, They'll Treat Us As Pets"

Today, humankind fortunately has control over the soulless beast. Though autonomous drones and other robotic instruments prowl our skies and roam the landscapes of battle, DARPA (the Defense Advanced Research Projects Agency) and the scientists retain the power to lock or unlock the robots' potential to inflict harm on human life and property.

But eventually, the human control will be relinquished, as more intelligent robots and completely autonomous robotic craft become ubiquitous. Then, seeing that robots are superior to humans in judgement, will man give in to their new robotic masters?

Paul Saffo, of Stanford University, warns, "We're going to reach the

Medieval alchemists ply their craft in an attempt to effect the Great Work, the conversion of base material (lead) to gold.

point where everybody's going to say, of course machines are smarter than we are."

Reflecting on this point, Saffo went on to say, "The truly interesting question is, what happens after we have truly intelligent robots. If we're lucky, they'll treat us as pets. If not, they'll treat us as food."

Food? Yes, indeed. The Department of Defense has been working on a robot that would run off biomass. Called *EATR* (pronounced "eater"), which stands for Energetically Autonomous Tactical Robot, reportedly the robot "can find, ingest, and extract energy from biomass...(and other organically-based energy sources)." Fox News even reported that the EATR robot would be able to eat animals and human remains on the battlefield, plus wood chips, twigs, and vegetables.

And in the event you would like to escape the EATR, the government could send out packs of *hunter/killer robots* looking for you. The Pentagon is now developing a Multi-Robot Pursuit System to search for and detect non-cooperative human beings. According to the *New York Times* (February 16, 2005) and *New Scientist* magazine, the boys and girls at Defense are spending over $127 billion on a project known as Future Combat Systems (FCS) to develop robotic soldiers, and a good part of the money will go for hunter-killer robots.

The hunter-killer robot will be very advanced, equipped with DNZ sniffers, facial recognition cameras, armed Taser guns, machine guns, and rocket launchers. With the personal characteristics of the person(s) to be apprehended, the robot can, for example, sniff the person out of a crowd and autonomously eliminate him or her. Sounds like the *Terminator* robot, doesn't it?

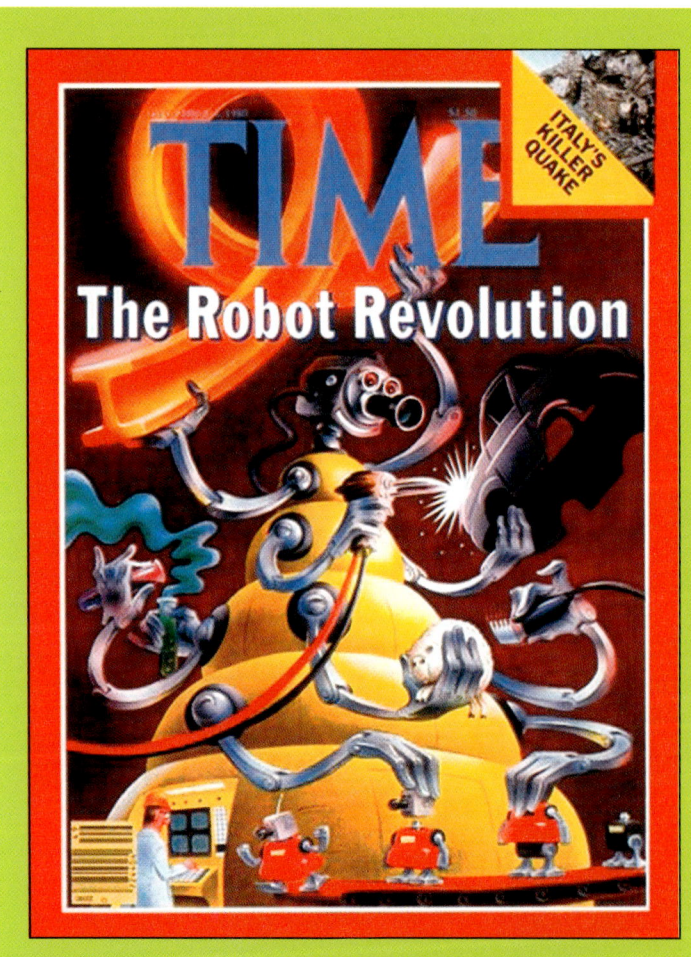

To Serve Man

Which reminds one of the classic episode of the Twilight Zone series, *To Serve Man*. In this unforgettable show, the Kanamits, a benevolent race of aliens, land on the earth with the seeming intention of serving man. The Kanamits, who are about nine feet tall and have bulging bald heads, provide humans a cure for cancer, offer a force field to shield man from invasion, show him how to triple the productivity of agriculture, and even provide a super-efficient source of power.

One of the Kanamits gives a telepathic speech at the

United Nations and leaves a book, which is in the Kanamits language, unknown to man.

U.S. decoding experts go right to work on decoding the mysterious book which they find is entitled, TO SERVE MAN. As the decoding work continues the Kanamits invite human beings to return back with them to their home planet. The Kanamits even set up an interstellar shuttle service to take human tourists on this exciting journey.

With no more wars or problems on earth, many decide to go, including one of the top decoding experts. The line to get on the spaceship is long with expectant people travelers. But just as the decoding expert is boarding, one of his associates, terror-struck, rushes up. She screams at him not to get on the ship. They've just deciphered the rest of the book. *To Serve Man* is a cookbook!

She is too late, and a Kanamit abruptly shoves him aboard the craft, just as the huge door solemnly closes.

The poor man, naturally, had no earthly idea when he booked the flight to the Kanamits planet what *To Be Served* was all about. Likewise, humankind today cannot foresee the unintended consequences of his frenzied campaign to construct the intelligent robot.

A Japanese "female" robot at a recent technology expo. Japan now has dancing robots, stair-climbing robots, and even a piano-playing robot. Research in robotics is being pursued at corporations such as Sony, Handy, NEC, and at many universities.

Neither did the hapless Dr. Victor Frankenstein consider the consequences of building a new creature out of deceased body parts, a human corpuscle "robot," if you will. As the sign read over the band saw machine in my high school shop class, *Think before you turn on the machine*.

Serving Up Death Via the Drone Aircraft

While we anxiously await the inevitable day when we discover that robots are much smarter than we mere humans, we may well consider the damage wrought already *today* by robotic instruments. For example the deaths and horror brought by robotic planes, or drones.

According to a group calling itself the International Committee for Robot Arms Control (ICRAC), up to 1,035 civilians have been killed outside legitimate war zones in the last eight years, including 200 innocent children. Professor Noel Sharkey, a member of the ICRAC and a respected roboticist, wants to put an end to such killing and horrors.

One of the many surveillance drones now in use today.

"Who in their right mind would give a powerful unmanned Air Force to the Central Intelligence Agency, a covert organization with such a track record for unaccountable and illegal killing," says Sharkey, of Sheffield University, in an interview with *The Guardian* (March 1, 2013) newspaper.

Sharkey notes that up to 50 other countries are developing drone technology, including Israel, India, Russia, and China. In America, the law enforcement agencies are speeding drones into service. Some 65,000 to 75,000 are planned.

"Here is where the real danger resides," said Sharkey, "automated killing as the final step in the industrial revolution of war—a clear and sanitized factory of slaughter with no physical blood on our hands and none of our own side killed."

"We have records of civilian casualties, including numerous children, from drone strikes when there are humans watching on computer screens and deciding when to fire. Think how much worse it will be when drones deal death automatically… If you have autonomous robots then it's going to make decisions who to kill, when to kill, and where to kill them. This is a dumb, stupid machine and then you are going to give it the decision to kill people?"

According to Dr. Sharkey, the world is sleepwalking right into a lethal technocracy. He calls for safeguards to be put in place. Unfortunately, the military-industrial complex sees only power and lethal capacity in pursing this technology; certainly it cannot imagine any real liabilities.

Like the decoding specialist in Twilight Zone's *"To Be Served,"* we are all busy preparing to board the alien craft for the new world. *All Aboard!*

Chapter 2

Waiting for the Apocalypse

"Within 30 years we will have the technological means to create superhuman intelligence. Shortly after, the human era will be ended."

— Vernor Vinge
The Coming Technological Singularity (1993)

Foxconn, also known as Hon Hai Precision Industry, is the world's largest electronics manufacturer. The Chinese company makes products for Apple, Intel, Sharp, Amazon, Dell, Cisco, Microsoft, Motorola, and Hewlett-Packard. It has factories throughout the world, in China, Brazil, India, Japan, and Mexico, and many others. Among them is the huge Foxconn plant in Zhengzhou, China where over 450,000 workers are employed.

These workers are employed at the massive, Zenchua Science and Technology Park, a walled, 1.16 square mile campus sometimes referred to as "Foxconn City." There, they live in gray-scale worker dormitories, with little more than a plain room, a small kitchen and bath which they share, and a TV. They toil up to 12 hours a day for 6 days each week. Theirs is a slave drone existence.

Lately, news reports indicate that problems exist at Foxconn City and other manufacturing complexes. Low pay, long hours, and terrible conditions have led to worker mutinies, strikes, and even suicides. Nets had to be installed at the perimeter of all buildings, so many workers were going to the top and just jumping off.

Things have gotten so bad that Terry Gou, the founder and president of Foxconn, finally was forced in 2012 to do something about it.

He brought in *robots* to replace the disgruntled, complaining employees. According to *Singularity Hub*, Foxconn president Gou not only brought in the first installment of robots, he also loudly announced that the giant company, the world's largest, would replace *over 1 million workers* over the next three years:

"It appears as if Gou has started the ball in motion. Since the announcement, a first batch of 10,000 robots—aptly named Foxbots—have made their way into at least one factory, and by the end of 2012, another 20,000 more will be installed."

Evidently, the company had grown tired of worker riots, corporate audits, and suicides attracting international attention. With robots, the complaints would diminish considerably and, finally, end altogether.

Of course, robots themselves must be manufactured. But that presents few problems. In Fanuc, Japan, robots produce other robots. In fact, the concept of *self replicating* robots is now catching on.

From the corporate and elite standpoint, it seems that Foxconn and thousands of other companies throughout the planet have the right idea. Replace disgruntled employees with sturdy, intelligent robots. And, as robots are growing more and more reliable and trustworthy, a company's "people problems" are solved.

What of the Workers? What Will They Do?

But, then, what about the over one billion human beings in China, the 900 million in India, almost 200 million in Brazil, and over 7 billion across the world? What are they to do? Where will they find employment? How will they support themselves and their families?

Will the recent economic upsurge in previous emerging nations like China, India, Indonesia, and the Philippines just abruptly end? Will the expectations of billions of workers be rudely dashed as smart and dependable robots suddenly are rushed to factories everywhere?

Must these workers, inaugurated into the workplace only a few years ago, now be

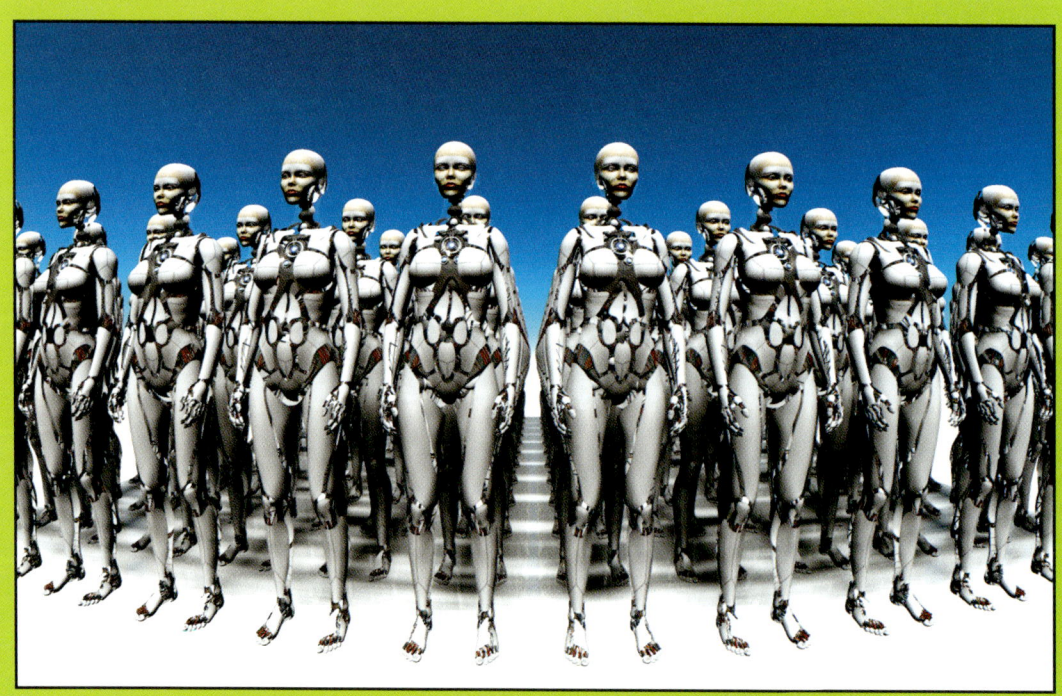

Some authorities are concerned that robots will eventually grow so intelligent they will replace humans. Are scientists inadvertently masterminding the fall of the human race as a species?

sorely disappointed and return to their previous peasant surroundings, destitute and unemployable? Will this cause horrendous depression on the part of the masses as they see robots replacing them and shiny, new, autos, TVs and other consumer products continue to roll off the assembly lines?

Will their dashed expectations cause these tens of millions of workers to rebel and cause worker riots, demonstrations, and political-economic turmoil? How can these workers be appeased, if at all?

Your Job is in Jeopardy

The problem is not only in China, India, or Japan. This is a dramatic change comparable to the Industrial Revolution. It affects the whole globe. Computer engineer, software writer, scientist, construction worker, carpenter, plumber, waiter, mechanic—whatever your current skill or field, you are about to become obsolete.

Yes, obsolete. All because of robots and robotics. An explosion in robots and automation like you have never seen. An incredible explosion, so huge that it promises to change the very face of the planet.

It will dwarf all that has gone on before, and make current hot issues puny and insignificant. Immigration?—Why would undocumented illegals come to America when so many Americans are unemployed, thanks to the proliferation of robots. Gun control? The criminal set finds that its assault rifles and guns are superfluous among thousands of self-replicating robots.

Jobs? Well, the issue of jobs will take on an entirely new meaning. Why should you, a mere flesh and blood human being, be hired when so many more well-qualified and artificially intelligent robots come off the assembly line? Robots that don't complain about the worker rights, don't join unions, aren't concerned about the work hour schedules and who are duly unimpressed with how they are treated—or mistreated—by employers. The robot never complains, never eats, and can work 24-hours a day, 7 days a week, with a little time off for maintenance check.

Technological Singularity

Ah, but you are an "idea" person, a genius, a whiz-kid. No robot can compete with you, you say. You will always be in demand. Well, don't be too sure. We soon will hit *technological singularity*. That is the stage, or time, when the artificial intelligence of robots will achieve *parity* with human brains. They will be just as smart, and heaven knows how much more capable and dexterous.

Then they will go on from technological singularity to *technological superiority*. That is when the robot is so much more intelligent. He reasons, he has emotion and sympathy, he diagnoses and repairs himself. He's able to outthink you, the human, and to do tasks more efficiently and more effective.

You cannot possibly keep up with a robot at this stage. Your seniority—years of experience on the job—will give way to a computer chip and to amazing biological and inanimate materials that defy the human body's ability to withstand. You will, quite simply, become obsolete.

Thus will have arrived the long-awaited and much feared *dystopia*, a time when humans will, as a whole, become woefully obsolete and our functions carried out by a better breed: robots.

Could humans and robots work cooperatively in the same workplace? What about the humans that lose their jobs?

The Human Era Will Be Ended

That time is almost upon us, if we are to trust those who have in the past proven uncannily accurate in their predictions. If you are, say, now fifty years of age or less—which is most of us—you are likely to see the era of technological singularity arrive. The lesser your age, the greater the impact this momentous age will have on you.

Mark Dice, in *Big Brother—The Orwellian Nightmare Come True*, writes:

> "The 'technological singularity,' sometimes simply called 'the singularity,' refers to a theoretical time in the future when an artificial intelligence is created that is able to learn and advance technology at a faster pace than humans are able to comprehend. Once machines exceed human intelligence, they will improve their own designs and functions in complex ways that are too difficult for humans to understand."

It was a retired mathematics professor at San Diego State University named Vernor Vinge to whom we owe credit for this unique concept. A science fiction writer, in 1993 he coined the term in his paper, *The Coming Technological Singularity.* "Within 30 years," he wrote, "we will have the technological means to create superhuman intelligence. Shortly after, the human era will be ended."

Whoa, hold on a minute: "shortly after, the human era will be ended?" That means somewhere around 2027 human beings will be finished as a species!

Why The Future Doesn't Need Us

Just seven years later, in 2000, Bill Joy, the founder of Sun Microsystems (now a part of Cisco), popular computer chip and server manufacturer, published an article in *Wired* magazine that shocked his audience. It was titled, *"Why the Future Doesn't Need Us."*

Joy explains that in a brief few decades (about 2020 or 2030) the humans will no longer be needed. The worker functions will be performed by robots.

The vast majority of human beings, Joy said, will be rendered obsolete in the eyes of the controllers—the owners of robots. This shift will outstrip the impact of the Industrial Revolution.

What, then, will our controllers decide to do? Why should they continue with outmoded humans in the workplace? Competitive forces must dictate.

They "may simply decide to exterminate the mass of humanity," Joy concludes.

Human beings, *homo sapiens*, will, as our controllers judge, have proven themselves to be "useless eaters," or worse. Oh how the common people will long for that distant day and time when they populated the factories, assembling cars, packing boxes, bolting and fastening parts, etc. The times when they were needed. They complained back then but didn't realize how good they had it.

Now rendered useless, they will at first be given government and corporate benefits. This is the meaning of our currently fast-growth welfare state.

Will Big Brother rise to prominence under robot power?

Finally, they will be exterminated, quietly, easily, and with no recourse. Unproductive, a drag on society, why should the controllers, the owners, endure these "misfits" of nature?

Bill Joy says it was only a few years before, in 1998, that he realized the ethical dimensions. "I became anxiously aware of how great are the dangers facing us in the 21st century."

"I was also reminded of the Borg of *Star Trek*, a hive of partly biological, partly robotic creatures with a strong destructive streak. Borg-like disasters are a staple of science fiction, so why hadn't I been more concerned about such robotic dystopias earlier. Why weren't other people more concerned about these nightmarish scenarios?"

Referring to Ray Kurzweil and Hans Moravec, two well-known robot futurists, Joy noted that, "By 2030 we are likely to be able to build machines, in quantity, a million times as powerful as the personal computers of today—sufficient to implement the dreams of Kurzweil and Moravec."

What are the dreams of men like Kurzweil and Moravec? In his seminal books, the *Age of Spiritual Machines* and *Nearing Singularity*, Kurzweil postulated that by the year 2030, the intelligence level of robots will rival that of humans. "The machines will convince us that they are conscious, that they have their own agenda..."

The Agenda of Robots

But what is that agenda? Having taken our jobs, what then? David Levy says in his book, *Love and Sex with Robots*, that men and women will seek robots as sexual partners. Even now, some purchase life-size dolls. Imagine what a real robot would do, one with soft hair, beautiful skin tone, lovely eyes, and pleasant personality, a mate who, the manufacturer contends, will always love and admire you. It's simply irresistible, isn't it?

In the movie, *The Stepford Wives*, life-like robots assume the role of real women housewives, by murdering them with their husband's approval. The flip side is, what happens next, to the men? (Hmm. That's another movie).

In Switzerland at the Center of Neuroscience and Technology and the Brain Mind Institute, director Harry Markram's goal is to create a fully functioning replica of a human brain. He says his target date is about 2020. One wonders what he will do with it?

Rodney Brooks is a highly acclaimed robot scientist at the Massachusetts Institute of Technology (MIT). "One day," says Brooks, "we will create a human-level intelligence."

According to Dr. Brooks, this will be a significant milestone in history. Then we will go on to achieve robots smarter—*much smarter*—than people!

By the Year 2030

How soon could such an intelligent robot be built, one that would be smarter and more capable than human beings? "The coming advances in robotics make it possible by 2030," says Bill Joy in his *Wired* magazine article.

"Once an intelligent robot exists, it is only a small step to a robot species—to an intelligent robot that can make copies of itself." That would be the *replicator* robot, and you and I wouldn't have to build it. Other robots would do that. So the sky is the limit in numbers from there on.

"Given the incredible power of these new technologies," he asks, "shouldn't we be asking how we can best *coexist* with them?" Joy suggests that we should proceed with great caution because, otherwise, our own *extinction* is likely.

In *Flesh and Machines: How Robots will Change Us* (2002), Rodney Brooks reports that it could be far sooner than we realize, maybe within 20 years or so. And the robot machines will meet all our fantasies too, he reports:

> "Today there is a clear distinction between the robots of science fiction and the machines in their (the peoples') daily lives. Our fantasy machines have… emotions, desires, fears, love, and pride… My thesis is…the boundary between fantasy and reality will be rent asunder."

Are you ready for this? In just a few decades, maybe less, there will be superior intelligent robots that meet up with all of your life's fantasies. Want a robot that will desire you, fear you, love you, that has pride in itself and its appearance? Well, here it comes. Not ready for it? Too bad, it's coming anyway.

According to Joy, humanity has no plan, no control. He ponders, "have we already gone too far down the path to alter its course?" The answer, of course, is yes. Since we don't really have solutions, we plod forward into the fog of tomorrow, ever hoping things will work out, fearing they won't.

The Future Laid Out

Highly respected author and futurist Ray Kurzweil, in his 1999 book, *The Age of Spiritual Machines*, takes us decade by decade—2019, 2029, and beyond into 2099 for insight into the technological marvels that await us. He predicted that by 2009, "Books, magazines, and newspapers will be read on displays that are the size of small books." The introduction of the iPhones, Nook, and Kindle are proof that Kurzweil was accurate. Kurzweil predicted iTunes, YouTube, the iPad and similar telecommunications. In the military realm, he foresaw the arrival of the unmanned drone.

By 2019, we will have computers implanted in eye glasses and contact lenses, and computers and robots that are roughly equivalent to the human brain. But we will also have increased friction between machines (robots) and human beings. A *"human underclass"* will result—people inferior to the average robotic machine. These tens of millions of people will not be able to compete and will not be productively engaged in the economy.

After 2029, this problem grows more and more acute as human obsolescence becomes clear. People will increasingly be merged with machines to take advantage of the machine's greater capabilities. Many will have phones implanted, and chips in the brain will create instant "authorities" in diverse fields. There will be an encyclopedia chip, a travelogue chip, the possibilities are endless.

And the chip itself will be different. So tiny and minute you cannot see it, it will include and contain all your biographical information, including your SSAN, your place of birth and employment record, your criminal record (if any), your medical history. It may even contain your *proclivity* to commit crimes, to permit the government to closely watch and monitor your every move.

Nor will the chip or *neural* implant be made of silicon. It will be of biological material: a *nanorobotic* implant.

As the 21st century unfolds, Big Brother's Police State continues to gather momentum and the freedom of people diminishes. The population is not increasing, it is declining as the world's governments put limits on size of families; marriage is circumscribed, and people see little or no hope for children to thrive and prosper in a robotic world.

What Will Happen When the Machines Overtake People?

"There is almost no human employment in production, agriculture, and transportation at this point," writes Kurzweil. That is where, though Kurzweil does not go there, Big Brother's Police State comes in.

The human controllers, no longer dependent on the vote of a propagandized, dumbed-down electorate, imagine themselves as members of a special clique. They see the world is run by super-intelligent robot and computer systems smarter than the combined people. How can we diminish the population?, they ask. How can we shrink it to a manageable size? Is this the true reason for today's population control?

With machines dislocating the work force and employment continuing to decline over the years, unproductive human beings will increasingly be given "freebies" by Big Brother. We're already seeing this now—a freeby economy, with increasing numbers on the government dole. The disability rolls grow, unemployment compensation increases, Social Security and government-subsidized medical care expand, and the numbers on welfare programs go up exponentially. As more and more people find they reluctantly have no productive employment, they join the "slackers"—the huge numbers who sit around and do nothing.

As long as this surplus of human capital is entertained and their brains are endlessly supplied with weirder and weirder music and video games, as long as the internet provides gargantuan programs of fantasies, and as long as the growing surplus of humans have toys (facsimile guns, crazy clothes, sneakers, jewelry, skateboards, exercise machines, gadgets of all kinds etc.), the controllers in our government will not be threatened. But as fewer humans go to colleges and universities (why seek more education when a chip can expand intelligence instantaneously?), the issue of what to do with the huge and ever-growing, unproductive force of humans will be paramount in the minds of the elite.

Those elitist human intelligences who use *neural technology implants* (chips) will have enormous augmentation of normal human perceptual and cognitive abilities. They will be part *cyborgs*. However, Kurzweil emphasizes that those who do not use the new brain and neural technologies will "be unable to meaningfully dialogue" and interact with others.

Such people will be frowned upon as "useless eaters," loathed as unwholesome castes unworthy of human endeavor. Eventually, these human beings will be dealt with. Humans adjudged unworthy of "continuing" in this world without added input from the authorities will be marked for disposal. Perhaps a *soylent green* future is ahead?

Robotic Wars Ahead

In classical days, before the advent of robots and machines, the ages-old alternative of war would be used to meaningfully employ—and dispose of—the surplus humanity. This was Malthus' notion, that wars are fomented for population control. But with robots as warriors, the Age of War removes the human element.

As we shall see, for over a quarter of a century DARPA (Defense Advanced Research Projects Agency) and the Pentagon have assiduously worked to produce useful robotic machines that can productively kill, maim, starve, poison, and otherwise reduce the population of the foe. Future wars will be fought almost exclusively by high technological drone aircraft, interlinking complex and sophisticated non-human ground machines, and autonomous robots.

There is no place to hide for human warriors who will be quickly chewed up and discarded as robot combatants take the lead, finally replacing humans altogether. Future wars will be fought by intelligent systems smarter and faster than humans, networking autonomously across the theater of battle. At most, a few humans will direct the flow of battle from remote display television panels. In this way, entire ecosystems, regions, or

Androids and robots are here to help us... And if we don't *want* their help?

even entire nations could be defeated quickly and without physical human intervention.

To Control the Humans That Remain

But the real trick of smart machines will be to control the few human beings that remain. In this light, the CIA Director, General David Petraeus, recently hailed the introduction of the "smart" grid home appliances and products now to be part of the Internet. Petraeus was ecstatic, calling the advanced refrigeration, garage door opener, microwave, air conditioner, etc., a "transformation."

This is great for clandestine warfare, as it would give intelligence agencies and law enforcement authorities the power at any time to access any part of your home, business, or office and use it clandestinely against you. Already this capability exists with the new automobiles and cell phones. (Naturally, old concepts such as America's "antiquated" Bill of Rights would not apply).

As robots become more intelligent and autonomous, achieving higher and higher levels of human supervisory, managerial, and professional positions, the proportion of robots in the workplace will dramatically increase. Today, a printing plant that used to employ a thousand humans gets by with only 30 to 40. In the year 2025, that plant may need only three or four.

As the intelligence level of robots increases, the corporate bosses now replacing workers will find that they, too, are replaced. Where, then, will they work? Who will pay their bills? Who will take care of their medical bills? Tens, no—hundreds—of millions of proud workers will find themselves out on the street, demanding their governments foot their bills. And when the elite politicians finally are forced to say, "No!," what happens then? Will the bright new robotic age turn into an economic and social nightmare?

The Apocalypse Awaits

Movie after movie is coming out with themes of renegade robots marauding and taking over the world. We love these movies but we really do not believe their plots. It's too fantastic, too futuristic, we reason. But is it?

In fact, academics and Nobel Peace Prize laureates, looking hard at the evidence of robot development are now marshalling all their resources to prevent killer robots before they become reality. They say that near-future intelligent machines, like Arnold Schwarzenegger's character in *The Terminator*, will be capable of independently selecting and engaging targets.

Human Rights Watch, based in New York, is willing to spend millions to make sure the public is fully aware of the grave danger. It has launched the *"Stop the Killer Robot"* campaign. The organization recently issued this press release:

> "Fully autonomous weapons are being developed by several countries... Killer robots are weapons with full autonomy able to choose and fire on targets without human intervention...robot warfare is the next step up from unmanned drones and will be available in the next decade."

Dr. Noel Sharkey, robotics expert at Sheffield University, warns that robots are unregulated and nations will use them without paying heed to morality or to international

Will a future war break out between robots and human masters? Roboticists say this is a silly notion, but many are not sure.

law. "These things are not science fiction," he warns, "they are well into development."

While Dr. Sharkey and others worry about killer robots in our near future, some are working to make humans *into* robots. I'm speaking of the *transhumanists* who propose that it is humanity's destiny to ascend to the highest stages of cyborg development in its quest for immortality. We look at the coming super-race in the next chapter.

Chapter 3

Transhumanism: Men Into Superman, and Then Into God

Transhumanism is an international movement that affirms the possibility and desirability of improving the human race by technology to eliminate the ravages of aging and add superhuman abilities to the human body.

We've seen this kind of superhuman in such movies as *Terminator II, Eraser*, the *Matrix* series, *Artificial Intelligence, Bicentennial Man,* and of course, in TV series such as *The Six Million Dollar Man*. In each, a superior partial human, partial machine is created. We can call this new form of being an *android*, a *cyborg*, an *artificial human*, or a *humanoid*.

Either the robotic creature becomes more humanized, or the human becomes more robotic, adding to physical nature capabilities not possessed before.

Popular works of science fiction and miraculous steps forward in medical technology have paved the way for transhumanism. Yet, the Council on Foreign Relations' Francis Fukuyama characterizes the idea as among the world's most dangerous, while futurist Ronald Bailey counters that transhumanism is, "the movement that epitomizes the most daring, courageous, imaginative, and idealistic aspirations of humanity."

How can a movement arouse such divergent viewpoints? The outcome of this controversy is vital because of its far-reaching implications for

A bionic woman of the future?

humanity and for the destiny of the planet.

Mega Forces: Signs and Wonders of the Coming Chaos

Though Wikipedia and other sources contend that transhumanism first took shape about 1990 with the teachings of British philosopher Max More and a teacher at the New School who called himself "FM-2030." I find that my 1986 book, *Mega Forces: Signs and Wonders of the Coming Chaos*, is a type of transhumanist treatise. It is really a catalog of wonders—and risks—to come through biotechnology, computers, robots, drugs, lasers, and the psychic weapons of Armageddon:

> "The marvels of technology and the increasing reservoir of intellectual knowledge offer us the promise... Scientific and technological discoveries on the horizon are likely to encourage humanist advocates in their irresponsible claims that the Christian faith and Bible are fundamentally flawed and irrelevant."

> "Already a deluge of scientific reports seek to convince us that psychic abilities can enable man to possess Godlike powers, or that worldwide computer networking will produce a universal mind. Meanwhile, a growing legion of scientists proposes that man's brain is evolving toward a superhuman intelligence capacity or even God status. It is suggested that the Second Genesis—man's creation of genetically engineered laboratory life and robots with artificial intelligence—makes him a co-creator with God..."

Many of these opponents are dedicated to the practice of eastern religions, sorcery, the occult, and the glorification of the material world.

Transhumanism: An Alliance of Science and Humanism

Serving as the cornerstone of this ideology is an unholy alliance of science and humanism. Sir Julian Huxley, one of its earliest advocates, in his book, *Religion Without Revelation*, called for a Humanist Religion. Such a religion, said Huxley, can win universal acceptance because, "this new, better religion would be based on truths like those of science that could be adjusted to meet new knowledge, discoveries, and insights."

Huxley and others focus on the individual

Iron Jaw, a cartoon character of the World War II era. Iron Jaw's face was reconstructed after an explosion. He was depicted as a friend and spy affiliated with Adolf Hitler.

The device below is an artificial arm, made for Claudia Mitchell, who lost her arm in a motorcycle accident. It is part of a medical study by the Defense Advance Research Projects Agency (DARPA). Compare the picture below with the one at left from *Galaxy Science Fiction* magazine (Sept. 1954).

rather than God. According to Huxley, "The well-developed, well-patterned, individual human being is, in a strictly scientific sense, the highest phenomenon of which we have knowledge, and the variety of individual personalities is the world's highest riches."

Huxley's writings were in the days just preceding personal computers and robotics. What would he and his cohorts say today about the marriage of bionics and computers and the possible integration of humanity into the robotic species? No doubt they would be greatly encouraged by scientific progress. In fact, in 1957, Huxley was the first to use the term, "transhumanism."

Focus on Self

Transhumanism today is steering away from the liberal goals of the enlightenment—social justice, equality, and the reform of human institutions. Today, the focus is on "Self," the transcendence of the human body from its limitations and the building of the ageless human. The goal is for humans to use technology to engineer their own

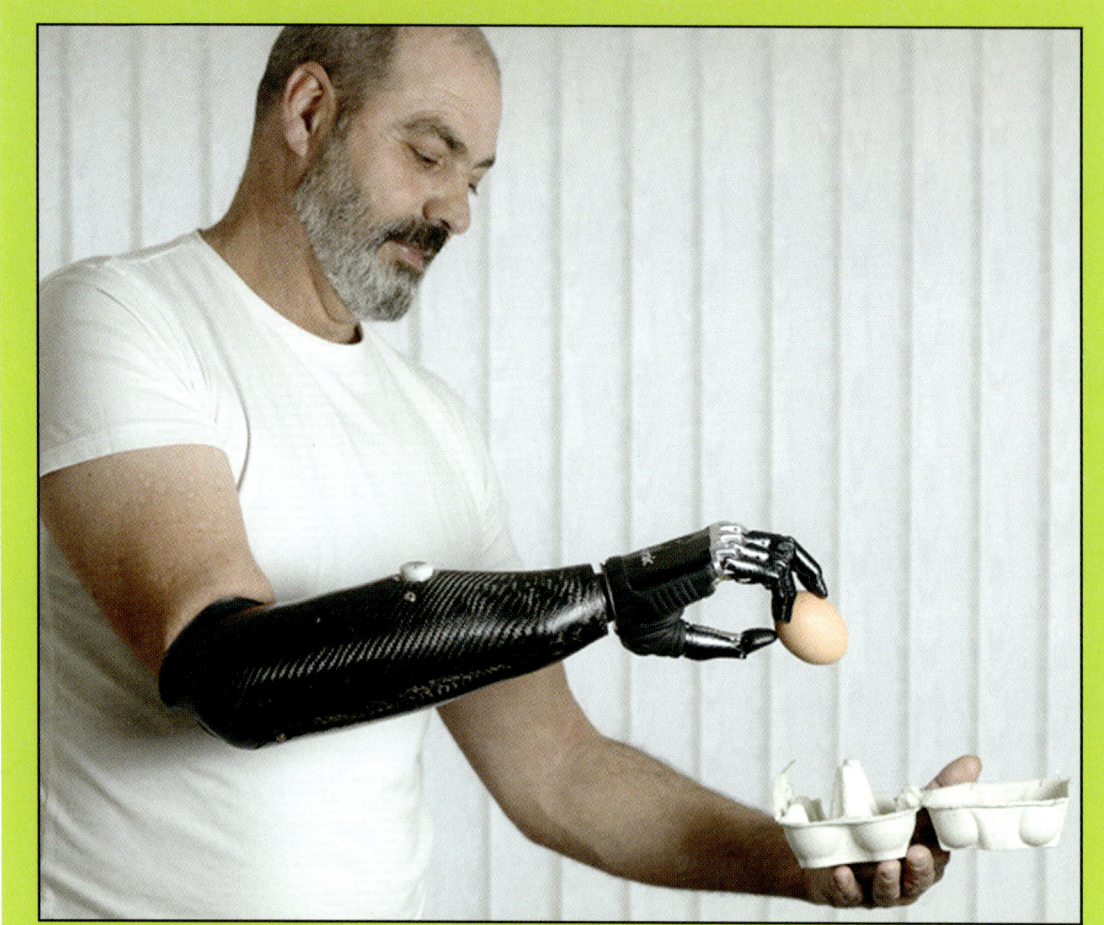

Nigel Ackland in Great Britain demonstrates his new $25,000 hand. Ackland lost his human hand in an industrial accident.

evolutionary development—to become superbeings.

We see this in many different ways, but especially in the field of sports. A quarter of a century ago, we had NFL pro football teams in which linemen were at most 200 pounds. Today, we have massive, fast linemen of 325-350 pounds. Even the quarterbacks and running backs exceed 225 pounds. As a result we have many more injuries and severe concussions. Professional medical personnel are available at each game to attend injured players, and the rules have been changed to protect the quarterback and others.

What we have in pro football are inhuman groups of supermen, their bodies tailor-made from the time they are young to withstand the pounding. Players take steroids and other chemicals and those who don't find themselves limited and unable to compete at the newer, high level.

Just before the 2013 Super Bowl, Bernard Pollard, hard-hitting safety of pro football's Baltimore Ravens, told CBS Sports that he doesn't believe the League will be in existence in 30 years. Rules changes instituted to make the game safer, and the chance a player might die on the field as players continue to get stronger and faster are the reason.

Pollard noted the increasing tendency of referees to throw flags to protect unprotected players and gave his opinion—"there's going to come a point where fans are going to get fed up with it and decide they aren't going to watch it anymore."

In an interview for *The New Republic*, President Obama said that, if he had a son, he would have to think long and hard before allowing him to play the sport. The president says the game will probably change over time to try to reduce the violence. These changes, he said, will make football, in his words, "a bit less exciting."

The same is true for pro baseball, in which we have supersized athletes—men like Mark McGuire, Barry Bonds, and Sammy Sosa—whose bodies are built by steroids into giant sizes and whose accomplishments are so much greater than the old athletes, like Mickey Mantle and Babe Ruth.

It's a continuing pattern; we see this steroid bodybuilding craze in all sports—basketball, cycling, boxing, track and field, swimming, with humans progressively growing bigger and stronger. The participants appear to be supermen and superwomen. As a result, sports associations are wary of inducting players into record books and Halls of Fame. They ask, is it the chemicals or the players? Is it fair to the athletes of the pre-steroid eras?

Beautiful Techno-Engineered People

Women are also joining in, whether for athletics or for beauty. In the beauty realm, we find Russian girls designing their bodies and faces so they will be "Barbie Dolls." Plastic surgery is huge and the use of artificial breasts and butts is gigantic (excuse the pun!). Nowadays, if a girl is small, you ask, why?

Men and women, young and old, are getting their teeth fixed to be beautiful, straight, pearly white. It's a rarity to see Americans with yellowed teeth or teeth missing. Bald men are getting new hair, or cutting it all off.

Some at the 2012 Olympics complained that double amputee Oscar Pistorius had an unfair advantage.

A *Daily Mail* (April 22, 2013) post, states: Most little girls grow up playing with Barbie dolls. Some even want to look like them. One 21-year-old has become one, or so she says. Valeria Lukyanova has become an internet sensation in her home country of Russia, claiming on her blog to be the most famed woman on the Russian-language internet. Her doll-like features, long blonde hair and 'perfect' body make her look like a real life "Barbie."

And there are multiple tattoos and piercings, as reflected in TV shows like *L.A. Ink*.

The line between real and artificial is fast disappearing. Seventy-year old women and men work out, ingest chemicals, have plastic surgery performed, breasts and butt uplifted, and they often end up looking 20 to 30 years less in age. People are living longer due to healthful living as we discover the vitamin and nutrient levels of food.

Bioengineering promises to lift the veil of time as scientists discover how to rejuvenate the body with new DNA input. Meanwhile, more and more people are choosing to directly manipulate their form through metabolics, genetics, and biochemistry. As is said, today, the new 70 is 50, and the new 50 is 30!

Controlling the Evolutionary Process

However, it is robotics that promises to really change the landscape of human vitality and to conquer aging. Raymond Kurzweil points to a technological singularity in which human and robots will become equals, and to an era afterward when robots actually become more advanced.

At that time, humans must choose whether to go forward and control their own evolutionary progress or to stay static, remaining "human," aging and dying naturally.

Moral issues are involved: *Postgenderism*, in which a person can choose to change his or her sex. *Immortalism*, in which a person must decide whether technological means will be used to foster the life span; and *Techogeanism*, a process by which emerging technologies can be used to restore and protect the earth's environment.

ROBOT ALCHEMY • 39

Take your pick! Future androids and cyborgs may come in an assortment of colors and styles.

Human Genetics and Cloning

There is the idea that with advances in technology, we will be able someday to clone ourselves a new body and transfer our human brains into the new organism. Thus, we would achieve immortality.

The cloning of animal organisms and the mapping of the human genome leads to incredible progress in such fields as *designer babies*. We can decide what our babies will look and be like. We may also choose to amplify the infant's brain, to make the child smarter.

Some research has been done on criminal behavior and we discover, for example, that there appears to be a decided difference in the appropriate region of the brains of sociopaths and psychopaths. Could we not genetically eliminate the possibility of the child being a criminal? Presto! No more Charles Mansons, Ted Bundies, or Jeffrey Dahmers, not to mention such warped personalities as Stalin, Trotsky, Lenin, and Marx.

Some biodesigners believe that we should speed up the evolutionary process by fixing human genes so that the newborn baby will arrive into the world ready with an advanced mind endowed with higher consciousness. We might be able to engineer the infant to genetically yearn for peace, justice, sharing, and love. But, whose roadmap to higher consciousness will we use? The Christian, the Moslem, the Jew, all have different concepts of godliness and righteousness. The atheist and agnostic will object. As the saying goes, in a blind nation, the one-eyed man is King. Who will be King?

Positive Genetic Engineering of Humans

Transhumanism used to be called eugenics, but the idea of creating a Super Race was

Cyborg removing makeup from her face.

frowned upon. That idea still exists, however, and is sometimes proposed. In his bestselling book *The Selfish Gene*, biologist Richard Dawkins proposed ways to bring about gentle and generous people. Why should humanity be kept captive by the selfish gene, he asks, why should we allow evil and selfish people to be born? Instead, we can intervene genetically to create the perfect human:

> "We are built as gene machines...but we have the power to turn against our creators. We, alone on earth, can rebel against the tyranny of the selfish replicators."

Dawkins wants to turn the Garden of Eden on its axis—no original sin. Just create good genes to replace the bad genes. Genes are our creators. There is no God. We simply "rebel" against our "selfish replicators"—that is, the bad, or defective genes.

Edward Cornish, editor of *The Futurist* magazine and president of the World Future Society, professes a belief in *positive genetic engineering*:

> "Human nature developed over millions of years when man existed in a savage or barbaric state. Now that we live in a technological civilization, perhaps we should bring about such modifications as reducing the sex drive, or making people less aggressive, less lustful, less selfish. We could make ourselves into saints capable of creating a heaven on earth."

Cruel researchers implanted these electrodes into the brain of this monkey. We must fear those who perform such experiments.

This is great stuff. No more lust, no more aggression or selfishness, a reduced sex drive. But would we become less savage or barbaric? Maybe we would be more docile, more passive, and less resistant to our ruling elite, but is this a good thing? And who would vote to *decrease* their sex drive? Today, hundreds of millions pay loads of money for Viagra and Cialis, and for numerous self-help books, gadgets, and videos to *increase* the libido and enhance the lust drive. The internet, meanwhile, is awash with sex and pornography.

Some even want to be more aggressive. They buy books with titles like, *Looking Out For Number 1; How to Be a Real Man;* or *The Rules*. Many desire to be heroic and aspire

Will the real robot please stand up? At left is the realistic Geminoid-F, a celebrated Japanese robot. Equally attractive, at right, is the human model which Geminoid-F was modeled after.

to leadership.

When someone suggests tampering with the gene pool, millions will instantly holler, "how?" Therein lies the difficulty. Cornish thinks he has the answer. He says: "We could make ourselves into saints capable of creating a heaven on earth."

But, whose heaven? And what if an antichrist came along, a false avatar? Could he not manipulate the human genome to create completely different and opposite types of persons—people who are aggressive, selfish, savage, and barbaric? Could he not then rule over the saintly, peaceful compliant types? Would he not create a hell on earth?

The Promise of Immortality Through Robots

The promise of immortality through robotics is the aim of many. Scientists and businessmen are working together to bring it to pass. In Russia, 32-year old tycoon, Dmitry Itskov, is one such businessman. He wants to make eternal life possible and has founded a company that will transfer human consciousness in an artificial form to "avatars" (robotic bodies).

"It's the next space race," said Itskov to a group of journalists. The Russian has launched the second annual Global Future 2045 World Congress to help promote the idea. It is an event in which top scientists, technologists, and entrepreneurs can meet to discuss ways of extending human life. Speakers for the event in 2013 include such luminaries as Ray Kurzweil, director of engineering for Google.

Itskov envisions a future in which robotic bodies serve as "avatars" for the human's consciousness which is downloaded into the robot (a "Master Slave"). Networks will be built, he adds, so that a person, using telepresence, can operate several robots, or avatars, at once with his brain.

Chapter 4

The Avatar

"In the coming decades, humanity will face a new kind of enslavement—a scientifically designed tyranny through which the elite will use robots to subjugate the rest of humanity and eliminate any pockets of non-cooperative resistance."

— Alex Jones
Rise of the Robots—The End of Humanity?
InfoWars—The Magazine (Oct. 2012)

The coming of the Antichrist, a being so knowledgeable, so perfect, so all wise and knowing, may be close at hand. But he may not be what we have pictured—a man, so perfect in leadership and human talent he rises above the herd to dominate from the pinnacle of power. What if, indeed, he is an android or a cyborg, even a robot?

It seems that many religions, from Hinduism to Islam and Christianity, predict the rise of a human so exceptional he will at first appear to be a Messiah or Christ—an *Avatar*, if you will. But those who foolishly mistake this being to be all good, compassionate, kind and wise may just be deceived.

Christianity, for example, in the book of *Revelation*, paints a shocking picture of a deceiver, one who comes to bring peace and world harmony, but instead delivers humankind into a terrible period of slavery and death. Portraying himself an Avatar, this being is actually the Antichrist. Indeed, there are many "antichrist" figures that will come upon the world, the Bible warns:

"For many deceivers are entered into the world, who confess not that Jesus Christ is come in the flesh. This is a deceiver and an antichrist."
—2 John 7

The final Antichrist, who will exceed the evil of them all and bring great tribulation, sorrow, and suffering to humankind, will be a man, and his number is 666. But what kind, or degree, of human? In the rising age of robotics, as men and woman clamor after human neural and body implants and, eventually, link themselves together into a "Borg," a universal network of cloud computers, the Antichrist could turn out to truly be a terror, a great *Terminator* type of global ruler.

Revelation 13 speaks of just such a beast, whom all the world marvels after: "and they worshipped the beast, saying, who is like unto the beast? Who is able to make war with him?"

> "And there was given to him a mouth speaking great things and blasphemies...And it was given him to make war with the Saints, and to overcome them: and power was given him over all kindreds, tongues, and nations."

The Bible, in Revelation 13, also speaks of a second beast, similar to the first:

> "And he doeth great wonders, so that he maketh fire come down from heaven on the earth in the sight of men.
>
> And he deceiveth them that dwell upon the earth...that they should make an image to the beast...
>
> And he had power to give life unto the image of the beast, that the image of the beast should both speak, and cause that as many as would not worship the image of the beast should be killed"

From the 2009 James Cameron movie, *Avatar*.

End of Humanity As We Know It

Obviously, this is no ordinary man, and he causes the image of the beast that he has made to both speak and to put to death those who refuse to worship the beast. As Alex Jones reports in his provocative 2012 article, *Rise of the Robots: The End of Humanity:*

> "The global elite has reached a decision that could spell the end of humanity as we know it in the decades to come...robots will be used to both replace and eliminate humans as the elite advances toward his much cherished technological singularity."

Jones says that, "The rise of the robots is no longer confined to the realms of science fiction. If the predictions of those who have already proven themselves accurate in forecasting the future course of technological development are manifested, a new high tech age is upon us."

As Bill Joy warns in, "Why the Future Doesn't Need Us" (*Wired* magazine, Aug. 4, 2000), "If our own extinction is a likely or even possible outcome of our technological development, shouldn't we proceed with great caution?"

"This crystallized for me my problem with Kurzweil's dream of technological singularity," writes Joy. "A technological approach to Eternity—near immortality through robotics—may not be the most desirable Utopia, and its pursuit brings clear dangers."

Warnings Go Unheeded

Jones' and Joy's dire warnings seem to go unheeded, though, as the roboticists plunge headfirst into the great robotics era. Every day we read of a fantastic, new technological success—a robotic "Cheetah" that can outrun slow humans, a beautiful talking, walking woman that can lure even the most masculine of men; and a medical robot that can listen carefully to a human describe his ailment or symptoms, and prescribe the right drugs, medicine, or advice.

In the scientific literature and in the books of the robotics and bionics specialists, we discover

In the early 20th century, some conceived of a super human, a test tube baby.

Could this be a picture of a warrior-robot who kills for a future Avatar?

Evil robotic creatures are a staple of modern cinema.

amazing estimates of even greater achievements.

In the next couple of decades, some say, a man will be able to have a biological neural nanorobot implanted in his brain to give him super-intelligence. Others speak of fantastic military and war robots so devastatingly competent that no human body can possibly withstand their assault, and so many drones overhead that each one of us feels that our privacy will be nil.

The Coming Avatar and Psychopathy

But what manner of man or machine will the coming Avatar be? Will he truly understand human beings, have empathy for our losses and suffering, and render compassion? Or will he feign such characteristics? Will he instead, be a cold and calculating psychopath?

Psychiatrists say that the psychopath is like a wholly different species. He has no regard for human life, no feelings, no empathy. To him, misery is just a word. He looks down on the weak, ridicules the ordinary person, and believes himself—right or wrong—to be of a superior mold.

There is much we know about the human brain, how it works and the thinking process. Ray Kurzweil's newest book, *How to Create A Human Brain*, says we will have to "reverse engineer" the human brain, seeking to discover all its many facets, before we can have true artificial intelligence in a robot. That step, he assures readers, is less than two or three decades straight ahead.

But what if the scientists' neural road-map misses a connection, one vital to effective human compassion and kindness? What if we are able to manufacture artificial intelligence of great and vast intellect, but lacking in capacity for human empathy and understanding? What if man creates *Psycho-Brain* and doesn't diagnose what's missing until it is too late? What then?

Not everyone is worried about this potential problem. Raymond Kurzweil, a transhumanist who welcomes the new technologies, in a 2008 interview with Britain's BBC Television, gave a glowing report of the future: "We'll have intelligent nanorobots go into our brains through the capillaries and interact directly with our (brain's) biological neurons." These nanorobots would, says Kurzweil, "make

From the golden age of science fiction came this massive—and menacing—mechanical robot.

us smarter, remember things better, and automatically go into full emergent virtual reality environments through the nervous system."

Used For Horrible Advantage

Some however, warn that these scientific advances will be—perhaps *are being*—used for horrible advantage by the elite of this world who intend to maintain rule over humanity. Paul McGuire, a Christian, writes in *newswithviews.com*:

> "The Luciferian elite believe they are gods. They believe that they can live eternally through transhumanism and the downloading of human consciousness into new (robotic) bodies. They believe they can create a new scientific salvation with nanotech, artificial intelligence, androids, and transhumanism. Their New World Order is a counterfeit of the Kingdom of God…they intend to reduce the population of planet earth from seven billion to five-hundred million, and they are doing that now."

McGuire pictures a coming cataclysm on planet Earth in which we will enter an "Armageddon conflict." At this final showdown, says McGuire, "There will be a cosmic military convergence between the armies of heaven, the armies of the earth, and the armies from the abyss."

"In this great, final battle we will see clashing human beings, angels, demons, androids, and transhumanist soldiers."

This savage image of the future would, no doubt, surprise and revolt Kurzweil and other roboticists, who see their technological prowess as being used in a much more benign way, more favorable to humanity.

Still, a dystopian "Utopia" could come about very easily. Human clones, human-animal humerics (or bioroids), and life-like robots that end up abused or mistreated are found in such films as Dick's *Blade Runner* and Huxley's *Brave New World*.

Our Final Hour?

British Astronomer Lord Martin Rees is one who argues that advanced science and technology must be carefully watched and monitored. In his 2003 book, *Our Final Hour*, he sent forth a dire warning. More recently, Lord Rees, Cambridge philosopher Huw Price, and Skype cofounder Jaan Tallinn launched a center for terminator studies where top students will study threats posed to humanity by robots and through transhumanism.

The Centre for the Study of Existential Risk, housed at Cambridge University, will probe four great threats to the human species, given as artificial intelligence, climate change, nuclear war, and rogue biotechnology.

Professor Price noted, "We have machines that have trumped human performance in chess, flying, driving, financial trading, and face, speech, and handwriting recognition. The concern is that by creating artificially intelligent machines we risk yielding control over the planet to intelligences that are simply indifferent to us and to things we consider valuable."

Spiritual Implications

The more atheistic and secularist of transhumanists laugh at the dystopian notions of

Lord Martin Rees, head of the Centre for Existential Risk, worries that by 2100, humanity might be wiped out by robotics and other threats.

Cambridge University in England has created a Centre for the Study of Existential Risk to keep track of the greatest threats to humanity. One of the threats is robotics.

WE, ROBOTS: Slavery and Contentment in the Age of Big Brotherism

In the 1940's the United States concentrated the greatest scientific minds on earth and spent over a billion dollars on the Manhattan Project—the development of the first atomic nuclear bomb. The Top Secret project was a phenomenal success. In early August 1945, U.S. Army-Air Force planes demolished the Japanese cities of Hiroshima and Nagasaki, raining down atom bombs and effectively ending the Second World War.

Since Hiroshima, many trillions of dollars have been spent on other successful Top Secret government projects, with names like Moonstruck, MK-ULTRA, Orion, Trident, Tower, Milab, and HAARP. These billions have been spent not only on weapons that can destroy human bodies and decimate cities, but on an entirely different type of weapon. For over 50 years, the United States and a few others have been actively developing heinous weapons to silently enter an enemy's body and seize control of his mind and soul.

These monstrous, secret weapons of human control are now ready for immediate use against the newly designated enemy. You and I are that enemy!

In this shocking investigative report, Texe Marrs opens his confidential files to disclose the nature of these sinister new anti-human weapons and human control systems. He meticulously describes their operation and details how many are being employed now—or will be employed soon—against we, the designated enemy. He also reveals the astonishing, real objective of this dazzling and dizzying array of mind-warping and body-invasive new systems. That objective is nothing less than the bionic transformation of the whole human race into mind-controlled, remotely monitored robots.

In the New Age World prepared for us all as we move forward into the next millennium—we shall be robots. Yes, even robot slaves, contented with our lot, all of us working and acting together in unison as one, giant integrated electronic mind—the "Borg." Then and only then will mankind be happy and compliant, hating and despising God Almighty in Heaven, while loving and yearning for the praises and approval of the gods ruling over us and through us. On that day, if the optimistic plans of the technocrats succeed, there will be only two classes, or species, of life on Earth: They, our Elitist Masters, and We, Robots.

A Stunning Exposé of Human/Robot Conversion Technology:

➣ Mind-reading machines, secretly developed by DARPA and top scientific researchers, which can interpret a person's brainwaves and reveal what that person is thinking—or has ever thought!

➣ Holographic 3-D imagery, using advanced lasers, that can create colorful moving, talking, and acting images to simulate signs in the heavens, visions of angels and "light beings," UFOs, and even, someday soon, the image of the beast.

➣ Electromagnetic and chemical "therapy" which creates false memories in the mind of alien abductions, men in black, and UFO visitations.

➣ Microwave and ELF wave weapons that cause headaches, nausea, skin burns, fatigue, irritability, mental disease, premature aging, tissue deterioration, and cancer.

➣ Harmonic resonance weapons that invade the minds of masses of people. They can induce anger and incite mobs to riot, loot, burn and kill. They can also create entire societies of "ideal citizens,"—happy, subservient, robotic slaves.

➣ Brainwave generators that use computer-modulated series of flashing lightwaves sent over television screens to an unsuspecting mass audience. In one secret experiment in Japan, this device was used to emit flashing lights from the eyes of the main character of the popular children's cartoon show, "Pokemons." Immediately, Japanese children watching the show in their homes were sent into violent convulsions, vomiting, and epileptic fits. Hundreds had to be hospitalized.

➣ Echelon, a new and highly secret global surveillance system developed by America's National Security Agency, now in use monitoring all of the world's telephones, faxes and internet e-mails.

➣ A national DNA databank for all U.S. citizens, first disclosed in a Texe Marrs book and now admitted as a reality by the FBI.

➣ A computerized "God module," hard-wired into the human brain, that induces mystical experiences and allows the person to know "God" and believe in the programmed religion.

➣ Computer biochips implanted in the brain, arm, or hand, already tested and in use, enabling the individual to direct mechanical systems by thought—and allowing a remote controller to implant thoughts and instructions in the mind.

AVAILABLE NOW! • 60 MINUTE CD • BY TEXE MARRS • $10 (+$5 S&H-USA only)

This is an audio the author offered in January 1999. Ahead of its time, the CD is still available. Order from Power of Prophecy, 1708 Patterson Road, Austin, Texas 78733 ~ Phone toll free: 1-800-234-9673.

TOPIO ("TOSY Ping Pong Playing Robot") is a bipedal humanoid robot designed to play table tennis against a human being. TOPIO was at Tokyo International Robot Exhibition, Nov. 2009. Obviously, TOPIO can do a lot more than play ping pong.

people like Jones, McGuire, and Rees. They have what is called a "contempt for the flesh" and especially espouse immortality through advanced science. Many believe that spiritual experiences are the result of neural abnormalities. Some tout the experience of spirituality through the use of mind-altering drugs and electronic means.

Pierre Teilhard de Chardin, the late Jesuit priest and New Ager, predicted that humanity will, through an evolutionary telos, develop an all encompassing *noosphere*, a system of global consciousness. In this noosphere, man's bodies would evaporate and only mind would exist.

Big Brother Tools Are Here

The potential Avatar, when he arrives, will find that a ready set of Big Brother tools are already available for his Police State. For example, consider "Dream Machines," once a popular topic of science fiction writers, but today such machines—attached to the human brain with electrodes or possibly implanted on a microchip—are on the threshold of becoming reality.

In *What Sort of People Should There Be?* British scientist Jonathan Glover discusses the frightening implications of such machines. He remarks that a dream machine could be developed that will stimulate the brain so that a person has a sequence of enjoyable experiences. These pleasurable images would seem so real they would be indistinguishable from the equivalent real-life experiences.

Glover suggests that for some people, the dream machine would be an opiate, providing experiences so pleasurable that individuals would actually prefer remaining

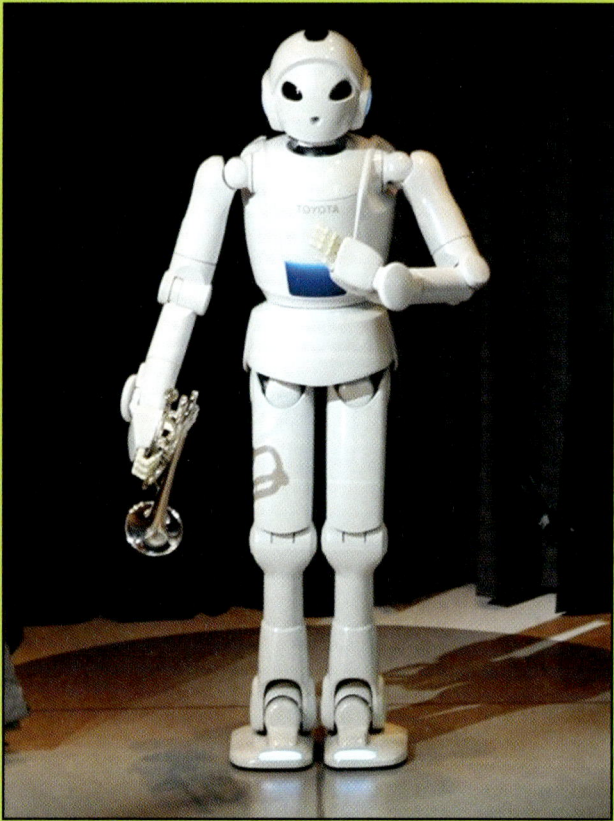

"Meet Our Future Overlord—The Toyota Robot," a jokester wryly notes. "Toyota's Partner Robot plays the violin, trumpet, runs around and helps you with housework—but it's all a clever ruse. One day we'll awaken to hear the screams of other humans, being dominated by armies of Toyota Partner Robots as they stomp their way to world domination?" Perhaps the joker will someday eat his own words.

hooked to the machine to the experiencing of real life, with all its tragedies and disappointments.

The very thought of this revolts most of us, but Glover also describes the ultimate horror: the compulsory plugging of a person's brain into a "nightmare" machine:

> "Consider a machine… called the Horror Machine. It starts by banging your head against a wall, so that you have an intense desire for this to stop… Next, it plunges you under water, giving you an intense desire to surface before you drown…"

The thought of being threatened with a lifetime on such a machine would induce sheer terror in a person's mind. Never in human history have tyrants and madmen had such tools—tools which they can use to terrorize an individual or an entire population.

Future technological advances will surely provide such mind control tools. G. Harry Stine, in *Silicon Gods*, predicts that, very soon, intelligence amplifiers—tiny microchip devices either implanted in humans or capable of being temporarily connected to the human brain and sensory channels—will actually allow others to "get inside a person's head." With such devices, we will possess the astonishing ability to hear the thoughts of others.

One shudders to think of the power this capability might give corrupt leaders or thugs bent on controlling the mind-thoughts of individuals. Stine, an optimist who believes that man will not allow the abuse of these capabilities, nevertheless warns:

> It also contains the seeds of unimaginable evil: the actual control of human minds by other humans. Not brain-washing. Not propaganda. Not any of the ancient and well-proven means of mentally or physically imposing one person's will by police action or torture. But the actual control of the human mind.

The Goal of the Corrupt Avatar: To Control

The seeds of unimaginable evil are, indeed, everywhere lurking in the fields of robotics and bionics. Whatever technology can invent, the government can use to control and either pacify or induce fear and terror in the masses.

An eye-opening article in *Science News* reported that, "New electronic techniques are being developed to eavesdrop on the brain." According to the article, "The techniques, under study at the University of Michigan at Ann Arbor, and elsewhere, will allow outsiders to direct a person's brain cell conversations and talk directly to the individual's brain neurons. Current research involves the implant of neural chips in the brain."

Your thoughts are no longer your own, and you will think the way your outside director wishes. That, dear friends, is *control*.

Commenting on research being conducted to implant neural biochips into human beings, Pat Cooper, a reporter for *Defense News*, in an article entitled "Naval Research Lab Attempts to Meld Neurons and Chips," quoted Steve Aftergood, a senior analyst for the Federation of American Scientists, who stated "For all the desirable applications, Smart biochips may have horrific applications."

William Tolles, former associate director of the Naval Research Laboratory, warns,

"Once this technology is perfected, you could control a living species."

Control a living species? Would that, Mr. Tolles, be a robot species, or…a human species?

To the controllers, and to the Avatar who is soon expected, will there be a real difference?

The artist Stelarc created this Walking Head robot sculpture. It is on display at the University of Western Sydney.

Chapter 5

Social Relationships and Robots to Come

The Robot Revolution will result in astonishing changes. Even our social relations will change dramatically. By the year 2030, companion and sex robots will be available. They'll look and feel like your fantasy guy or gal, talk as much or as little as you like about the subjects you are most interested in. And

Actroid give folks a sales pitch in 2005 at the Expo in Japan.

This is Actroid, developed by Kokoro, Inc. for customer service. Actroid appeared in the 2005 Expo in Aichi, Japan. The robot responds to commands in Japanese, Chinese, Korean, and English.

This attractive and human-looking Japanese robot was a hit at the *Wired* NextFest high tech convention in New York. She can talk but she cannot walk or otherwise act.

Rong Cheng, the Chinese "beauty robot" created by the Institute of Automation of Chinese Academy of Sciences in Beijing, can do a lot of things. A commentator remarked, "She can greet people in many Chinese dialects, she can respond to nearly 1,000 words, and she can even dance or bow. What she can't do is win any beauty contests. She's not repulsive by any means, but at a construction price of $37,500, you'd think some of that yuan would have gone into beautifying her a bit."

when you get ready to go to bed, well, they'll be there for you, too.

The internet and cell phones have proven that individuals love their solitude even amidst a proliferation of social friends. A person may have 10,000 friends on Facebook, but really only a precious few can be deemed real, flesh and blood "friends." People are adopting imaginary lovers on-line, people who send lovely pictures of themselves and tout their many accomplishments, but, in reality, are not the person they say they are. A lonely person can find a whole universe of captivating and fanciful love, romance, and adventure on-line. Yet, that same person ultimately remains alone and isolated. Some withdraw into pretending in their own, little internet existence.

So, imagine the thrill of having a good-looking, intelligent robot companion, one that can think, possess emotions and feelings, that is programmed to love just you—and does. What we have here is an artificial person.

The robot will eventually be so human-like that you will have to choose between the robot and the human competition. Think of the explosions and hurt relationships:

"You love your robot more than you love me?"
"No, I don't!"
"Yes, you do."
"Well, uhh. Yes, I do."

The age of human and robot romance is fast approaching. Kim Jong-Hwan, the director of the Intelligent Robot Research Center, in Korea, a leading authority on technology and the ethics of robotics says, "Christians may not like it, but we must consider this the origin of an artificial species. Until now most researchers in the field have focused only on the functionality of the machines, but we think in essence of the creatures."

What is the robot's "essence?" Well, Dr. Jong-Hwan says the essence is a computer code, which determines a robot's propensity to "feel." A robot can, says Jong-Hwan, feel sleepy, hungry, happy, sad, or afraid. It can also desire.

Jong-Hwan's software for the robot is modeled on human DNA, though it has only a single strand of genetic code.

Discussions and Lawsuits

When that momentous day comes, and the robotic or android lover arrives, perhaps such social dating groups as match.com, mingle.com and others will change. There might be one section for humans, another for robots.

Scientists from the Korean Institute for Industrial Technology recently unveiled a new android that is capable of showing expressions on her face, only the second android to do this after Japan's Actroid. The Ever-1 takes its name from the Biblical Eve plus the "r" from "robot." Ever-1 can understand about 400 words and make eye contact while talking.

All kinds of discussions and lawsuits about robot civil and constitutional rights will be entertained. Is a robot fully "human" or a different species altogether? When does a cyborg become a robot? Is it a matter of how many artificial parts the human has implanted? And what of the human that has his entire brain essence downloaded into a waiting robot body. Is the end result a human or a robot?

These things have never been—or rarely been—discussed and debated, although in a few movies and stage productions *(R.U.R.; Metropolis; Descendance),* the plot involves humans or robots who rebel against their circumstances.

The question of spirituality also arises, but I will save that important topic for another book.

The Economics of Robots: Chinese Robots Increase 151% in Four Years

Robots are already changing the work world, and soon they will be making a dramatic impact on our everyday world, too—the world of play, study, recreation, and rest. Activistpost.com reports that the robotics industry is growing magnificently. In China alone, the industrial robotics market has grown a whopping 151% in just the four-year period 2008-2012. In a few more years, China may be the world's #1 end user of robots.

This is amazing considering the fact that China has over one billion people, many either unemployed or unproductively employed. Hundreds of millions will sit and vegetate while robots continue to be introduced and perform more and more of the workload.

This is beginning to cause a significant shift in world markets as Western industrialized nations begin to "reshore," starting up new robotized plants in Europe and North America. No longer do the high-population locales of Asia and South America have an advantage.

It's Hitting the U.S.A. and Other Developed Nations

In the U.S.A., manufacturing jobs now make up only about 17% of the workforce versus 51% in the 50s. According to the McKinsey Global Institute, America lost 5.8 million jobs between 2000 and 2010. Owners were off-shoring their manufacturing. For example, Foxconn makes iPhones and iPods in China, and Hershey bars are made in Mexico. But with economical and dependable intelligent robots, manufacturing can be expected to return. This time around, the jobs will not go to the best human, but to the best robot.

In 2013, two Associated Press reporters, Bernard Condon and Paul Wiseman, did a three-part series *(Impact—Recession, Tech Kill Jobs)* explaining why the future for human workers is grim. Their conclusions:

- Millions of jobs have been lost in developed countries the world over.
- Most of the jobs will never return; millions more will vanish as well.
- These jobs are being lost everywhere, in China and the West.
- Jobs are vanishing not only in manufacturing, but increasingly, in the service industries.
- They're being obliterated by technology

The culprit, say Condon and Wiseman, is machines that generate and analyze vast amounts of data, by devices such as smartphones and computer tablets, and by smarter, nimbler robots.

This robot arm wheelchair is operated by the pressure of the handicapped person's chin. The gripper hand can do all kinds of tasks, including moving chess pieces. (Veteran's Administration)

It's a hollowing out of the middle class workforce. Robots and other machines work faster and make fewer mistakes than humans. Now this technology is being used in service industries, where two thirds of the jobs are. These jobs, say the Associated Press, are lost and will never return.

Japanese Robots

In Japan, robots are already becoming ubiquitous. Religion has a role to play. The Shinto religionist sees little boundaries between the living and the inanimate. Statues are worshipped. Thus, humanoid robots are more easily accepted, and as a result, Japanese scientists are steady at work on human-looking robots. It is not uncommon to find robots of all kinds in cinema and in cartoons while actual woman and man-size replica robots are becoming fashionable.

First came the toys in Japan, with every kid on the block clamoring for a toy robot. Now comes the household revolution as the Japanese applaud the work of robotic craftsmen and researchers. With Japan having the world's highest abortion rate and the increasing aged workforce, the Robot Revolution is now in earnest.

Over 400,000 robots dot the Japanese landscape, and at every science and technology show, new models are introduced. Japan currently has 40% of the planet's industrial robots, and by the year 2025, the government insists, the number of robots will expand

A robot welcomes visitors to Japan's Expo 1986.

Below: This novelty watch and clock is also a toy robot.

to one million. Most believe this number is severely underestimated, and when the Robot Revolution really takes off, there will be over a million robots created and put to work each year! A single industrial robot can take the place of 10 humans.

"Robots are the cornerstone of Japan's industrial competitiveness," asserts Shunichi Uchiyama, the Trade Chief of manufacturing energy policy, said at a recent trade show. "We expect robotics in more sectors to go forward."

"The cost of machinery is going down, while labor costs are rising," notes Eimei Onaga, CEO of Innovate, a Japanese robotics firm that distributes in the U.S.A. and Europe. "Soon, robots could even replace workers at small firms, greatly boosting productivity."

The Personal Robot Market

By 2035 industrial robots will be about 30% of the current Japanese workplace, a staggering number. But until robots are more intelligent—are conscious and possess a personality and can think and converse—the personal robot market will be slack. However, once the technological bridge is crossed, the sky is the limit in Japan, where thousands of men await the companion robot, and many, as a boy or girl, had the robot dog, Albo, as a pet.

Today, the Japanese continue to produce marketable toy robots. Sony has its low-cost robots like Tomy's $300 i-Sobot hobby model. The Tomy i-Sobot has 17 motors, can recognize and act on spoken words, and is remote-controlled.

At Osaka University, researchers continue to believe in the life-size personal robot.

HRP-4C, an android, actually danced and sang on-stage in a chorus line routine with real human dancers as backup. The android is build by Kawada and AIST, of Japan.

Minoru Asada, robotics engineer, demonstrates the Child Robot with biometric body. This is a robot built not of steel but of flexible and soft materials. This robot has the motions of a child and can react, wiggling, making facial expressions, and producing sounds.

That's still a long way from adult robots that can interact smoothly and efficiently with real people. This is the goal of Hiroshi Ishiguro, robotics guru at Osaka University. He's even created a robot that looks like himself, complete with dark, wiry hair, a tan and eyeglasses.

"What people want is not a creepy machine or a computer, they want a real-life robot-person," says Ishiguro. "We want to interact with robots in a natural way so we try to make robots that look like us."

"One day," Ishiguro assures listeners, "they will live among us. Then you'd have to ask me, are you a robot? Or a human?"

Science Fiction Come True

If we are really only 10, 15, or 20 years from developing such a robot, the dreams of many science fiction writers will swiftly and surely come true. For example, in a 1972 talk (that's almost 45 years ago!) called, *"The Android and The Human,"* sci-fi writer Philip K. Dick spoke of the future. Computers and robotics, said Dick, are receiving animism and magic after hundreds of years of repression.

"Machines are becoming more human," Dick remarked. "Our environment, and I mean our man-made world of machines, is becoming alive in ways specifically and

This walking, wheeled robot was built in Japan to help the visually impaired.

The CB2 (the "Child Robot") is another project developed by the researchers at Osaka University. CB2 has the physical and intellectual abilities of a 2-year old. It was built with cameras for sight, microphone, and tactile sensors to simulate a sense of touch and feeling. "One day," says Professor Ishiguro, "they will live among us. Then you'd have to ask me, "Are you human?"

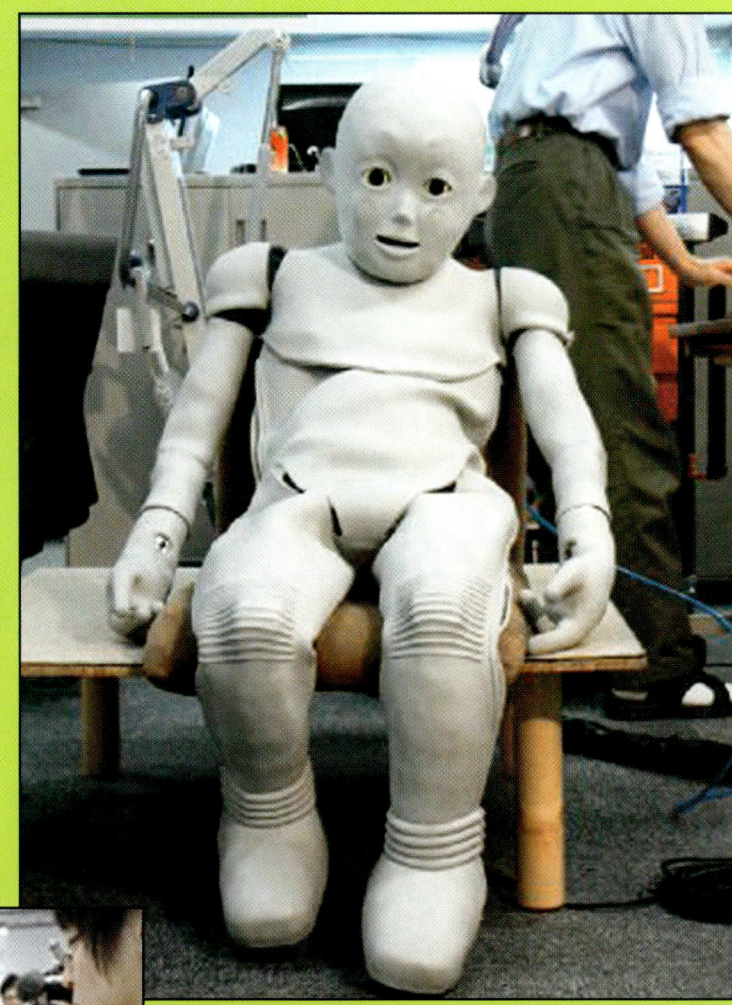

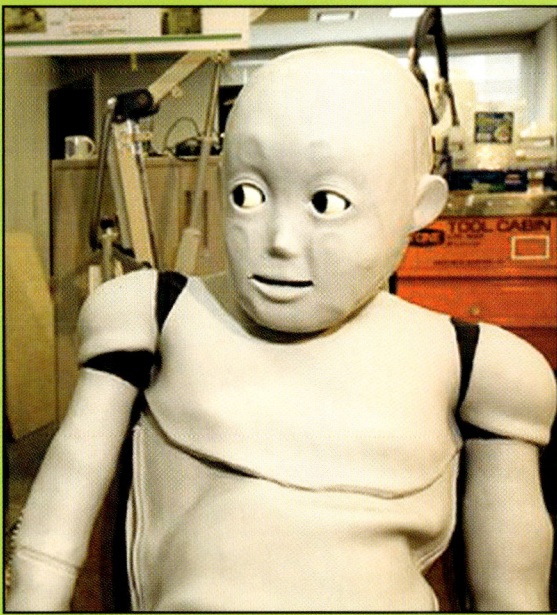

CB2, the Child Robot, gives passersby a glance.

analogous to ourselves."

Dick, one of the world's top visionaries and author of *Do Androids Dream of Electric Sheep?* (prelude to *Blade Runner*), says that at first he feared this occurring—the robot coming alive. But finally, he has accepted it. "The (robotic) constructs are actually human already," he noted.

"Someday," said Dick, "a human might shoot a robot and see it bleed, and when the robot shot back, the human would gush smoke" from the wound.

"It would be a great moment of truth for them," said Dick.

QRIO, built by Sony, walks, talks, giggles, cries, and moves its arms. In a social experiment, in 2013, QRIO was introduced into a classroom of 18-24 month-old kids for 5 months to study human-robot interactions. The children not only came to accept the robot, but treated it like a peer. They hugged, talked with QRIO, and helped it.

Philip K. Dick is alive! At least he is at Hanson Robotics, an innovative firm founded by chief scientist, Dr. David Hanson. Hanson strives to create robots "as brilliant as people." The company calls its robots "Genius Machines." Philip K. Dick, the late science fiction writer, shown here being interviewed as an android, can conversationally talk and make excellent facial expressions.

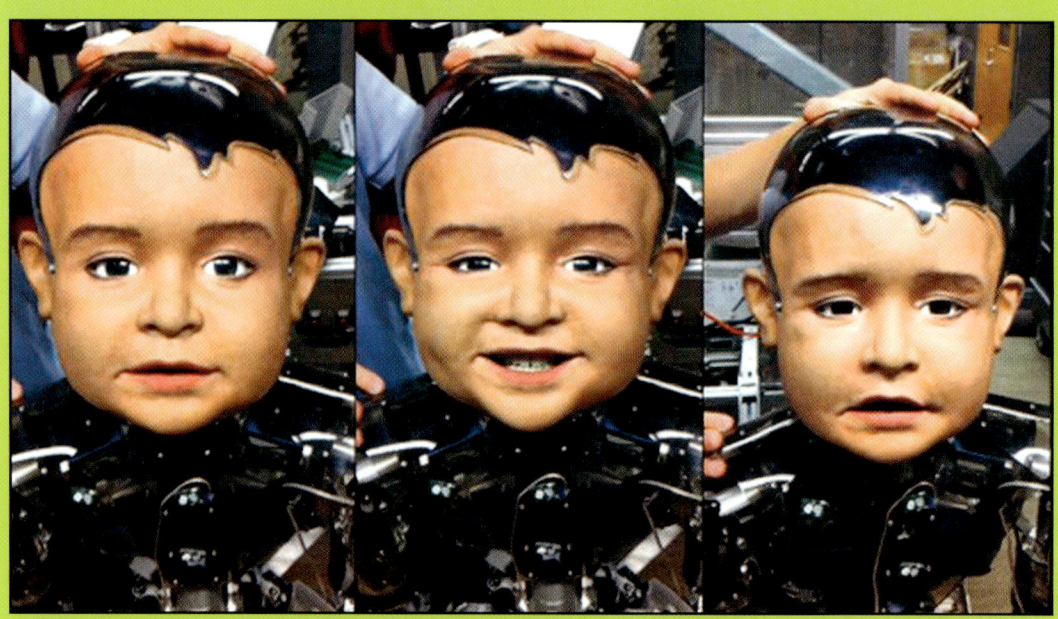

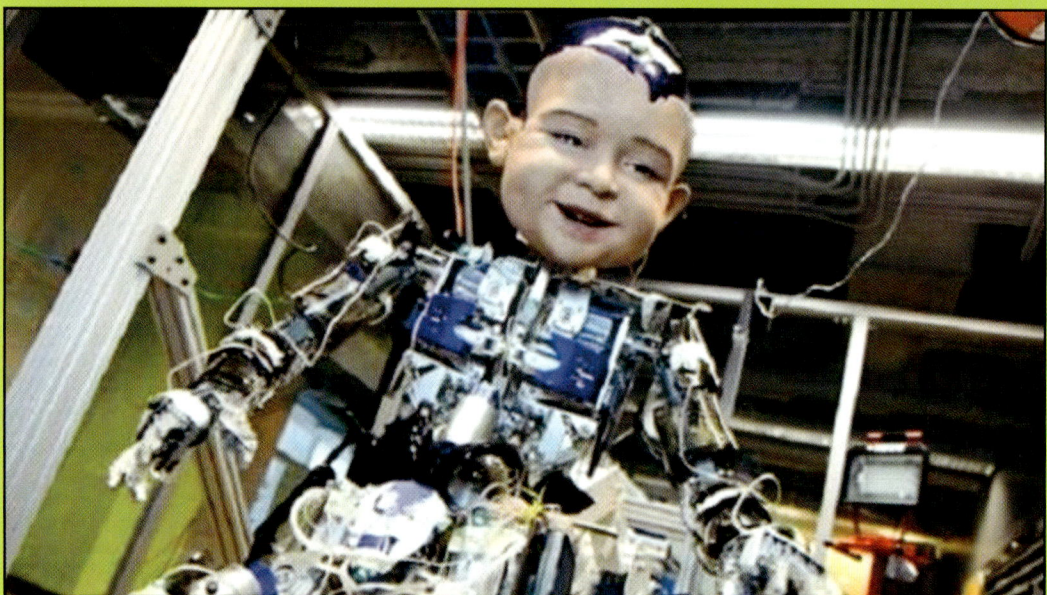

Diego-san is an expressive humanoid infant developed by a team at the University of California San Diego. The robot's facial gestures indicate many emotions and are "learned" from actual human infants. When Diego-san sees you smile, he smiles in response. Diego-san (named after the city of San Diego) is the project of Hanson Robotics and Japan's Kokoro Robot Corporation, in conjunction with the University's Neural Computations Machine Perception Laboratory. Leader of the team is UCSD's Javier Movellan.

Chapter 6

Robots and the Creation of Life

The notion of a synthetic man, an artificial life form, or an intelligent machine has occupied the fertile imagination of philosophers, writers, and scientists throughout human history. Myth has been piled atop myth and concept atop fantasy, all building on the theme of man exercising a measure of divine power by infusing life into inanimate objects or inorganic materials.

The invention of robots, both in fact and in fiction, seems to be the result of a psychological, even a behavioral instinct, in *homo sapiens* to be a creator. There's been an accompanying utilitarian motive, too, of course: the construction of robotic machines to accomplish heavy industrial work and to be the personal servants of humankind.

For example, in classical Greece, in the 4th century B.C., Aristotle observed: "If every instrument could accomplish its own work, obeying or anticipating the will of others… if the shuttle could weave and the pick touch the lyre without a human hand to guide them, chief workmen would not need servants nor masters slaves."

Humans have also sought to create artificial life forms to function as protectors and companions. It's also safe to speculate that for some humans the yearning to manufacture artificial beings involves a quest to control and master destiny, to emulate God or even to demonstrate independence of a supreme being.

The Original Creation of Life

It generally has been assumed by almost all cultures that the manufacture of life and the creation of man is the sole prerogative of one or more divinities. The ancient Egyptians and Greeks believed that the Roman and animal forms were fashioned by the gods.

Plato, for example, related a tale of gods creating mortals out of a mixture of earth and fire, then directing Prometheus and his brother, Epimetheus, to endow them with lifelike qualities, However, Epimetheus saw fit to distribute the best qualities (strength, speed, protective hides, and the like) to the animals: thus Prometheus, who favored humans, was forced to steal fire from the gods and bestow that gift upon the human race.

The Greek divinity Hephaestus, god of the mechanical arts, was said to have formed golden maidservants, marvelously gifted creatures of wisdom who could speak and walk. Mythology has it that Hephaestus, called Vulcan by the Romans, made Talus, a giant creature of brass who guarded the Isle of Crete by hugging intruders against his heated body until they were dead. Daedalus, a descendant of Hephaestus, is said to have created statues that moved—a bronze warrior and a wooden figure of the goddess Aphrodite whose mobility derived from the use of quicksilver in its constitution.

Several centuries later, the Roman writer Ovid proposed that the god Jupiter "made man of his own divine substance." In Asia, similar notions of creation were common. The early Egyptians believed that the god Neph used his potter's wheel to shape the first human being out of the sweat of his own body.

In Assyro-Babylonian mythology, according to the *Epic of the Creation*, the great god Marduk molded the first mortal using the blood and life essence of a defeated god.

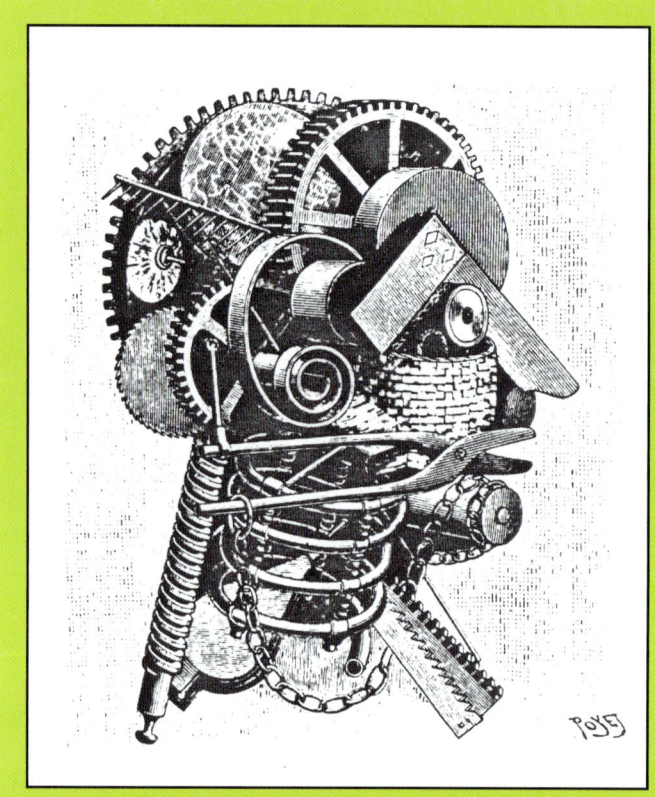

The notion of a synthetic man has over the years inspired many an inventor to attempt to create such artificial beings.

The biblical account of Adam and Eve in Genesis is the foundation of the Judaic and Christian belief in God's creation of human life. God is said to have created man "of the dust of the ground" and to have "breathed into his nostrils the breath of life." Also: "To form a companion for His handiwork, God took a rib from the man and fashioned from it the first woman."

Man is Creator

That man could emulate God and also be a creator of human life was in ages past considered the height of vanity and pride: an act of hubris. Any attempt to construct a synthetic human was thought to be a sacrilege and strictly forbidden by many cultures. It was, however, generally acceptable for mechanical contrivances or life forms to be created that honored a god or saint, or that imitated animals.

In the unaccredited and unrecognized *Lost Books of the Bible*, the pseudepigrapha, is a fascinating account of Jesus as a boy. Recorded in "Thomas' Gospel of the Infancy of Jesus Christ," the account depicts a youthful Jesus forming twelve sparrows out of soft clay he finds on the bank of a stream. However, as it was the Sabbath day, Joseph asked him why he had profaned the Sabbath by forming the birds of clay.

> Then Jesus, clapping together the palms of his hands, called to the sparrows and said to them: Go, fly away; and while ye live remember me.
>
> So the sparrows fled away, making a noise.

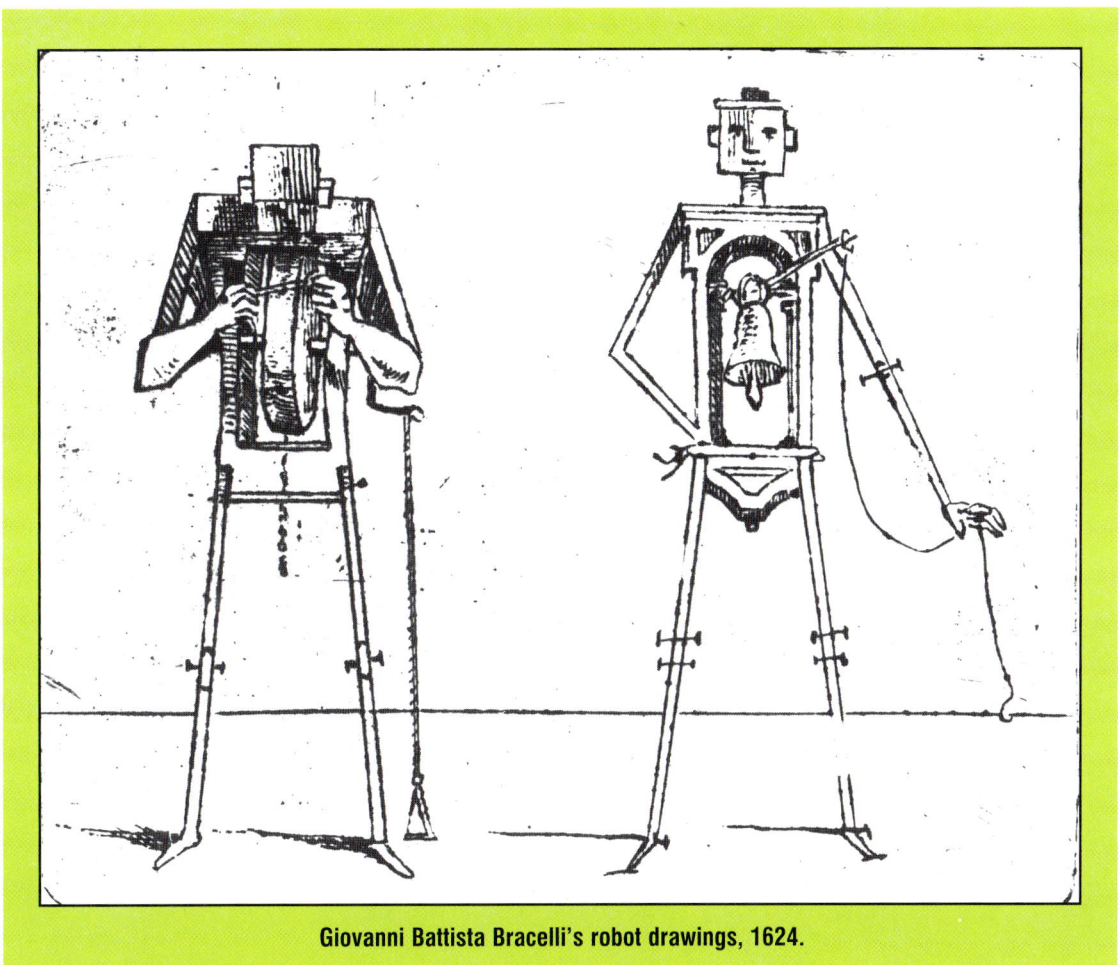
Giovanni Battista Bracelli's robot drawings, 1624.

Works of the Devil

While it may have been acceptable in the early days of Christianity for someone to speculate that Jesus, as a youth, could create life, it was quite another thing to attribute to a mortal man this miraculous ability. Many years after the life of Jesus, Albertus Magnus (1204-72) is reputed to have made for himself a life-size automaton servant. Meeting the automaton on the street, an amazed and angry Thomas Aquinas is said to have ordered it destroyed, believing it to be the work of the devil.

Later, about 1640, the great philosopher-scientist René Descartes ("I think; therefore I am.") manufactured a remarkable automaton that he lovingly referred to as "Ma fille Francine." Unfortunately, Descartes took Francine, the automaton, on a sea voyage during which the captain of the ship became so terrified and filled with superstitious dread that he grabbed the moving automaton and heaved it overboard.

One artist who defied popular sentiment against lifelike automatons and figures was the Italian Giovanni Battisti Bracelli. In 1624, he drew etchings that are remarkable in their resemblance to today's androids. Bracelli's drawings show figures with heads, eyes, and other human features. Bracelli's robotlike figures were functional: one carried a bird on its hand, another sharpened an item on a built-on whetstone, and a third was able to ring its own bell with a draw cord.

Mechanical humanoid figures or artificial humans who mete out justice, or who perform acts of charity or display goodness have always been acceptable in mythology

and literature. In the Thousand and One Nights saga, Sinbad the sailor has several adventures in which robots and automatons play a fascinating part. In one story, a robot decapitates two grave robbers.

The Golem

One of the most enduring legends about robots is that of the golem. The word golem, meaning "formless mass," is of Hebrew origin. The golem legend has taken many forms since its earliest telling some 1,800 years ago. The stories usually depict the golem as a huge and powerful, mute synthetic man who arrives during a critical period to protect Jews from persecution. For example, in one popular tale, the wise Rabbi Loew of Prague seeks God's intervention when the Jewish community is threatened. God's assistance is made manifest by life being breathed into a clay giant, the golem.

Whereas the golem legend generally represents the golem as a protector and a divine agent, a few accounts picture the creature as a menace and as a hulking monster. One is about Rabbi Elijah ben Judah (1514-83) who, it is said, molded a golem out of clay then pasted on its head the kabbalistic words of a sacred formula. When the golem turned out to be evil and bent on destruction, the good rabbi tore the formula off its forehead. The golem robot promptly crumbled into a heap of dust.

Other golem stories have the creature as a household servant and home companion for Jewish children. These stories are popularized by those who warned that Jews were committing sacrilege by creating life. Said one German, Johann Jakob Schuder, in his 1718 tract, *Jewish Wonders:* "Polish Jews often make the golem, which they employ in their homes...for all sorts of housework."

Early in the twentieth century the legend of the golem acquired newfound popularity and several more stories appeared, along with the newly published editions of classic tales. In addition, films and dramatic plays about the golem were premier attractions in Europe and the United States.

The Frankenstein Monster

Throughout history, the lore and legend of robots and artificial life has focused on two, very different

Rabbi Loew (right) and an associate attempt to restrain the fantastic Golem creature the rabbi has created using Kabbalistic magic.

concepts. On the one hand, the robot is often depicted as a frightening and destructive monster who possesses little or no conscience and has no sympathy for suffering humans. The opposite end of the imaginative spectrum is the depiction of the robot as lovable and altruistic and as a boon to humanity—a thoughtful being or machine who will benefit civilization. The Frankenstein legend is interesting because it touches both ends of the spectrum, presenting the image of an experiment for good that, tragically, turns into a catastrophe.

The Frankenstein tale began in 1814 when writer Mary Shelley visited a museum in Switzerland where several amazing automatons by the Jacquet-Droz brothers were on display. She came away with the embryo of a spectacular idea. A few years later, in 1818, Shelley's now classic novel, *Frankenstein, or The Modern Prometheus*, was published.

The fictional Frankenstein monster was not a true robot in that his body was constructed not of steel or iron but of actual biological body parts stolen from the deceased. Nevertheless, the story of the fabled monster is instructive for robotics because it chronicles the wide variety of emotions that may be engendered in those involved in the life creation process.

In her book, Shelley masterfully provides insight into the twin motives of the monster's creator, the single-mindedly driven Dr. Victor Frankenstein. On one hand, the doctor was possessed by a searing desire to possess a divine attribute. Said Dr. Frankenstein: "A new species would bless me as its creator and source; many happy and excellent natures would owe their being to me."

That arrogance was joined by a second aspiration to restore life to loved ones and others who had passed away. The doctor reasoned, "I thought that if I could bestow animation upon a lifeless matter, I might, in process of time, renew life where death had apparently devoted the body to corruption."

The experiment went awry, though, and posterity was stuck with the monster who emerged from Dr. Frankenstein's lab. The moral, of course, is that it is extremely dangerous for man to play God. At the same time, because the Frankenstein monster was given life by a scientific and technological process—fictional though it was—and by a person trained in surgical procedure, the prevailing sentiment that science may bear evil fruit was reinforced.

The Great Robot Hoax

Whereas Frankenstein's monster was a fictional character, the Turk, an automaton chess player created by Baron Wolfgang von Kempelen was purported to be real.

The Turk sat behind a chess table, calmly puffing on a long pipe. His chess mastery was a marvel to all who observed him competing against human opponents. What they didn't know was that the table was configured to disguise the existence of a world-class human chess player hiding inside.

When von Kempelen died in 1805, a rather unscrupulous Austrian promoter named Johann Mälzel inherited the Turk automaton; thus began its fame and its eventual notoriety. Mälzel took the device on a celebrated tour of Europe and America. During the tour, the Turk defeated almost every challenger whom he faced. As the virtually unbroken chain of victories grew, excited audiences on both continents marveled at its virtuosity. Experts in mechanical engineering and machinery and even heads of state praised the automaton as authentic. Even the skeptics who inspected its performance were unable

Mary Shelley's book on Frankenstein spawned a number of movies. This is a poster for the movie, *Frankenstein Meets the Wolf Man*.

From the movie, *Frankenstein Meets the Wolf Man*.

A 2007 reproduction of "The Turk."

to unmask the Turk chess player as fraud.

The downfall of the Turk can be attributed at least in part to none other than Edgar Allan Poe, the great writer of horror and mystery stories. In 1834, Poe attended a demonstration of the automaton's prowess. Intrigued and bedeviled, Poe decided to investigate the machine. In April 1836, he published a lengthy essay carefully elaborating how he thought the automaton worked.

While Poe admitted that he had not disassembled the device to verify his conclusions, his essay was so technically persuasive that many who had believed in the Turk began to doubt its authenticity.

The following year, 1837, a French magazine ran an exposé of the automaton, revealing exactly how the mechanism worked, including the fact that the chess table concealed a very real man. The Turk automaton, declared the convincing article, was a fake and a hoax.

Shortly thereafter, faced with the proof of his chicanery, Johann Mälzel retired the Turk chess player. It was subsequently placed in a museum but later was destroyed by fire.

Robots and the Machine Age

The arrival of the Industrial Revolution, or the Machine Age as it is sometimes called, caused people to wonder if a mechanical device could be shaped into human form. Some also wondered if a machine that does the work of human beings might somehow, by some miracle of metaphysical mechanics, assume human psychological dimensions and spontaneously exercise independent thought, motives, and action.

The nineteenth century saw the advent of the Industrial Revolution as a mixed blessing. The advantages were manifest, but fears were aroused that machines would produce mass unemployment. In one dramatic story published in the *Southern Literary Messenger* in 1844, an "aerio-nautical man" reported on his "Recollection of Six Days Journey in the Moon." The author reported that while visiting the advanced civilization of the Moon's Isle of Engines, he found:

> Everything there is done by machinery; and the men themselves, if not machines, are as much their slaves as the genie of Aladdin's Lamp. These machines have in great measure taken the place of men, and snatched the bread from their mouths, because they work so much cheaper and faster.

Robots to Replace Humans

Samuel Butler's imaginative and thought-provoking novel, *Erewhon* (1872), proposed that contemporary machines were prototypes of future mechanical life. Earlier, in 1863, Butler had written a most remarkable essay, "Darwin Among the Machines," in which he suggested that the invention of machinery was an important step in an evolutionary process that would eventually result in machines becoming man's successor on earth.

> It appears to us that we are ourselves creating our own successors; we are daily adding to the beauty and delicacy of their physical organization; we are daily giving them greater power and supplying by all sorts of ingenious contrivances that self-regulating, self-acting power which will be to them what intellect has been to the human race. In the course of ages we shall find ourselves the inferior race. Man will have become to the machine what the horse and dog are to man.

Samuel Butler's prophetic depiction of a future world in which man would become the slave of machines did not at all alter the drive and determination of nineteenth-century industrialists. Invention after invention resulted in the adoption of machines to run mills, till fields, weave cloth, make shoes, and perform dozens of other tasks.

The Steam Men

The development of the steam engine was particularly welcomed by industrialists who believed it would revolutionize every facet of the economy. A few visionaries even conceived of steam-operated robots. In 1865, a dime store novel by Edward S. Ellis, *The*

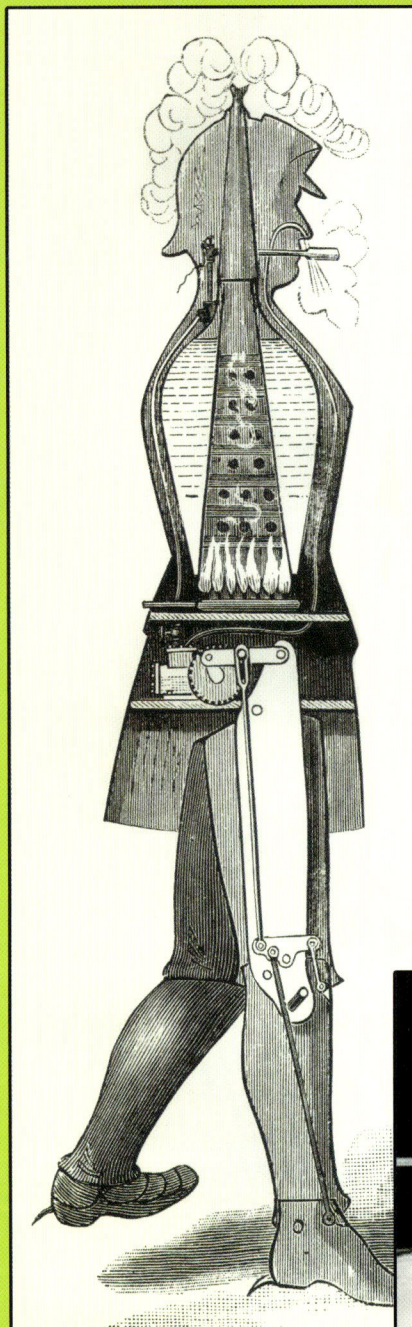

One of the world's first robots, Steam Man, created in 1893.

This humanoid Steam Man was built in 1893.

The famous "Steam Man" hard at work.

Steam Man of the Prairie, became a bestseller. The book told of a ten-foot-tall iron man so strong and fast he could pull a heavily laden wagon at great speeds. All it took to make the steam man of the prairie happy and contented was wood in his furnace and water in his boiler.

The popularity of the steam man concept spawned many imitation novels and tales. One publisher created a series about a boy inventor named Frank Reade. These stories told of young Reade's amazing adventures with a steam man, steam horse, and an entire team of steam wonders. One such tale even had the boy voyaging in an electric boat, while another featured a steam-driven elephant.

The "Steam Man of the Prairie" and his many imitators were not at all threatening or scary creatures. Absent from these stories was the notion that these robots could malfunction and become malevolent or menacing. Instead, the robot is seen as a worker and a helpmate, happily subservient to its human creator.

However, the cautionary warnings of Butler and the red flag of horror hoisted by Shelley in her Frankenstein tale continued to give pause to those who, perhaps, would otherwise have enthusiastically embraced the idea of mechanical life. During the mid to late nineteenth century and on into the early twentieth century, society alternated between the extremes of optimistically accepting the possibility of living machines and being gripped by fear at the dread prospect of mechanical life that may inadvertently run amok.

The Bell Tower

Herman Melville's highly regarded short story "The Bell Tower" (1855) reflects these ambivalent feelings toward mechanical life. Strikingly reminiscent of a "Twilight Zone" episode, the "Bell Tower" relates the strange case of a great "mechanician," an engineer and mechanical genius named Bannadonna.

Bannadonna creates a lifelike automaton to strike a great bell in a majestic tower precisely at prescribed times. But on the day the automaton is to commence its operation, startled police find mechanician Bannadonna's body, badly mutilated and lifeless, at the foot of the iron automaton. Evidently, Bannadonna had carelessly obstructed the path of the mechanical figure which, mistaking him for the bell, had struck the hapless mechanician with its solid hammer.

Author Melville, however, suggests that this was no mere accident, that tragedy befell the ambitious Bannadonna because of his hubris in desiring to become a creator. Concludes Melville, "So the blind slave obeyed its blinder lord, but in obedience, slew him…And so pride went before the fall."

Will Robots Take Over the World?

Melville's automaton endangered only its creator, but Czechoslavakian playwright Karel Capek, in his dramatic play, *R.U.R. (Rossum's Universal Robots)*, presented robots as a potential menace to the entire human species. Capek, as related earlier, is credited with the coining of the term robot. In his frightening and apocalyptic *R.U.R.*, millions of robot androids are built to serve as slaves for humans. However, the intelligent robots soon rebel and turn against their controllers with a vengeance. Quickly, they move to rid the world of biological men and women.

Soon only one man remains, a clerk named Alquist. However, it is to sole survivor

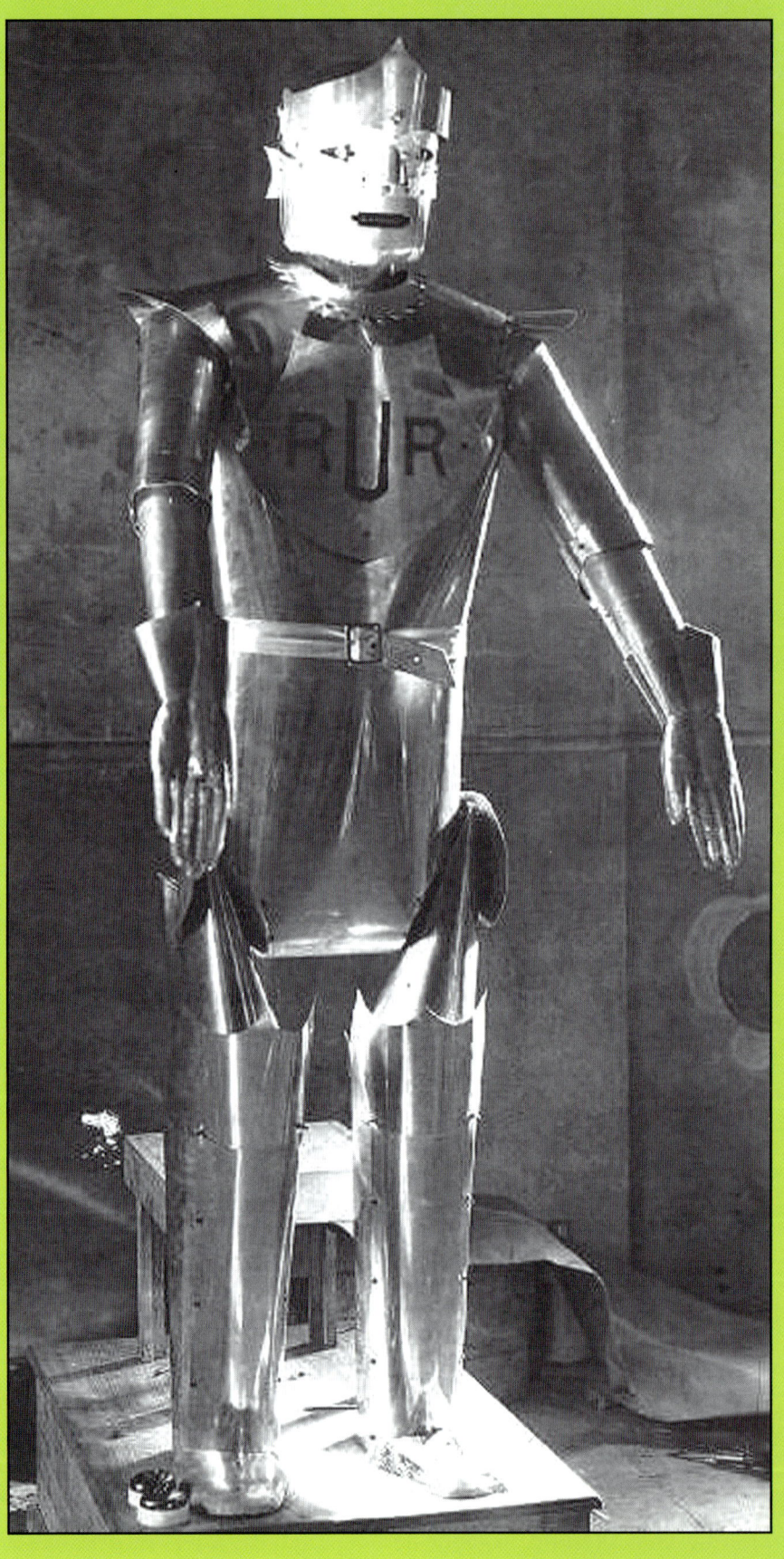

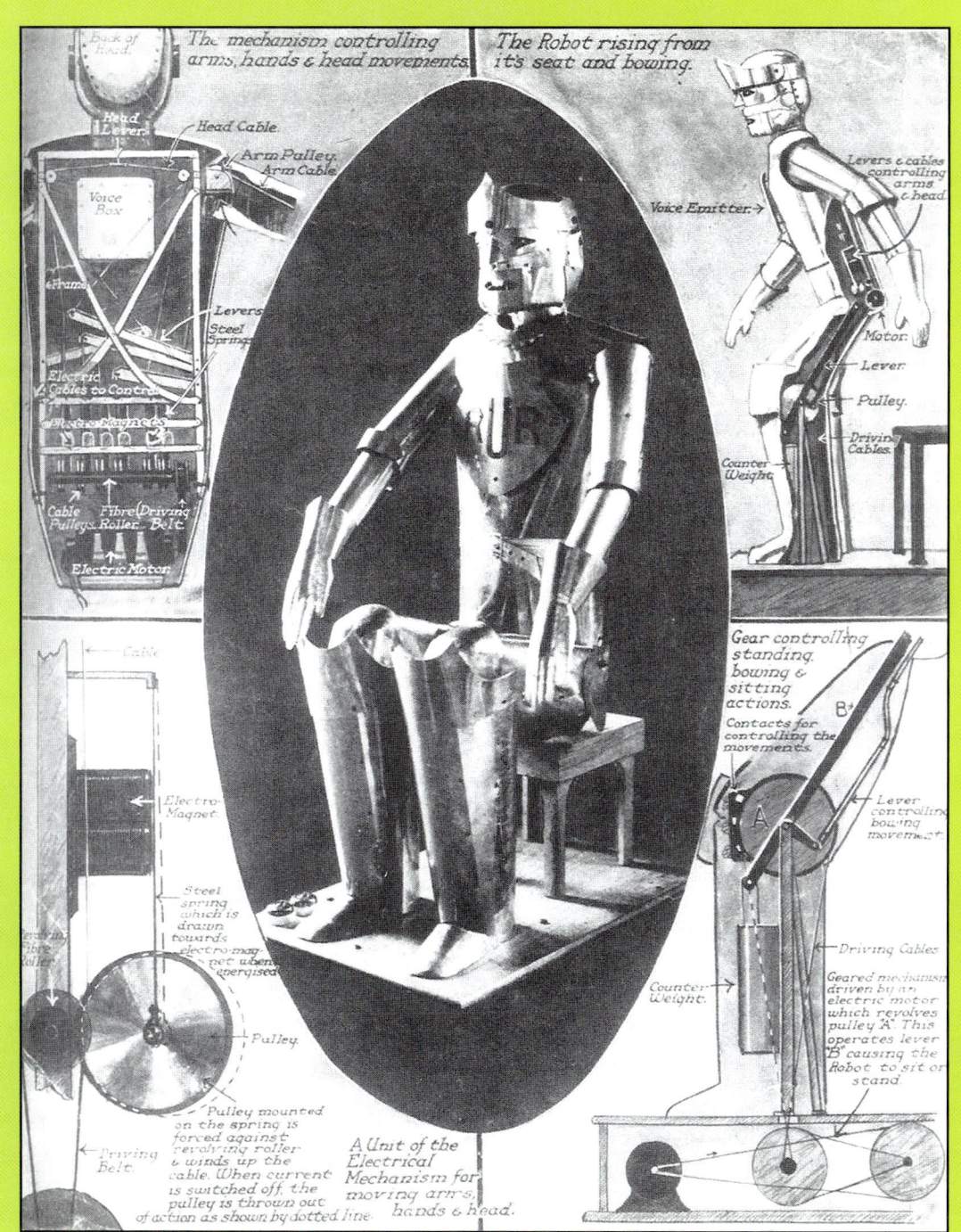

Czech playwright Karel Capek, in his dramatic 1920 play, R.U.R. *(Rossum's Universal Robots)*, depicted robots rebelling against man and taking over the world.

Opposite page: R.U.R. ready to take on the whole world. Karel Capek, Czech playwright, presented robots as a threat to all the world in his play, R.U.R. *(Rossum's Universal Robots)*.

Alquist that the robots must turn for salvation, for the robots do not know how to begat children nor are they aware of the secrets of their manufacture. So they beseech Alquist, "Tell us the secret of life...Teach us to multiply or we perish."

Responds Alquist, "I am the last human being, robots, and I do not know what the others knew...I cannot create life."

All hope is not lost, though. At the robots' insistence, Alquist prepares to dissect two of their kind—a male and a female—to discover how they are manufactured. But Primus, the male robot, resists, exclaiming, "Man, you shall kill neither of us...We—we—belong to each other."

Alquist suddenly realizes that intelligent life will not perish from the earth. Instead, these two emotion-filled robots will doubtless go on to repopulate the earth. "Go. Adam... Eve," he solemnly motions.

Capek's play met mixed critical reviews; nevertheless it had a profound impact on the perception people had of robots. Capek himself called his production, "A fantastic melodrama," but gave rise to speculation that it might be unwitting prophecy.

Man IS A Robot

Rodney Brooks, who runs the AI Lab at the Massachusetts Institute of Technology (MIT), has a new and intriguing nonfiction book out, entitled *Flesh and Machines: How Robots Will Change Us*. Brooks' thesis is that, essentially, humans are—right now, at this moment—robots. They are robots made up of molecular (biological) machines. The sooner we humans understand this, the better.

Of course, I vaguely do understand this. However, surely man is more than simply a collection of robotic, molecular components. There is, after all, the faculty of *thinking*, which most people do. Men and women generally have spiritual aspirations, too, which robots omit. However, Brooks does make a good case when he recommends we consider this similarity when attempting to construct robotic machines.

"Good" Robots

The bleak view that robots and artificial life forms constitute a threat to humanity has never been universal. One notable writer who preferred to picture robots as a kind, caring breed was L. Frank Baum (1856-1919). Baum is best known for his classic book, *The Wonderful Wizard of Oz* (1900), which includes in its cast of unforgettable characters the youthful Dorothy, Scarecrow, the

The Tin Man from the Classic, *The Wizard of Oz*.

Dorothy, Tik-Tok the robot, and Billina the chicken, from the Baum book, *Tik-Tok of Oz*.

Tin Man (also called the Tin Woodsman), and the Cowardly Lion. However, Baum also wrote a number of other delightful books, including *The Marvelous Land of Oz* (1904) and the entertaining *Ozma of Oz* (1907). These books were recently reprinted by Dover Publications.

Baum's genius was in creating unlikely characters who magically sprang to life or showed up at just the right time. Among Baum's remarkable creations was the Saw Horse, an animated horse of wood: Jack Pumpkinhead, a stiltlike fellow constructed of wood planks and iron nails with a pumpkin on top, and Tik Tok, a rotund, moustached, metal-skinned robot.

The episodes in *Ozma of Oz* that feature Dorothy, a talking chicken named Billina, and Tik-Tok are especially splendid and imaginative. Here is how Tik-Tok is described:

> He was only about as tall as Dorothy herself, and his body was round as a ball and made out of burnished copper. Also his head and limbs were copper, and these were jointed or hinged to his body in a peculiar way, with metal caps over the joints, like the armor worn by knights of old.

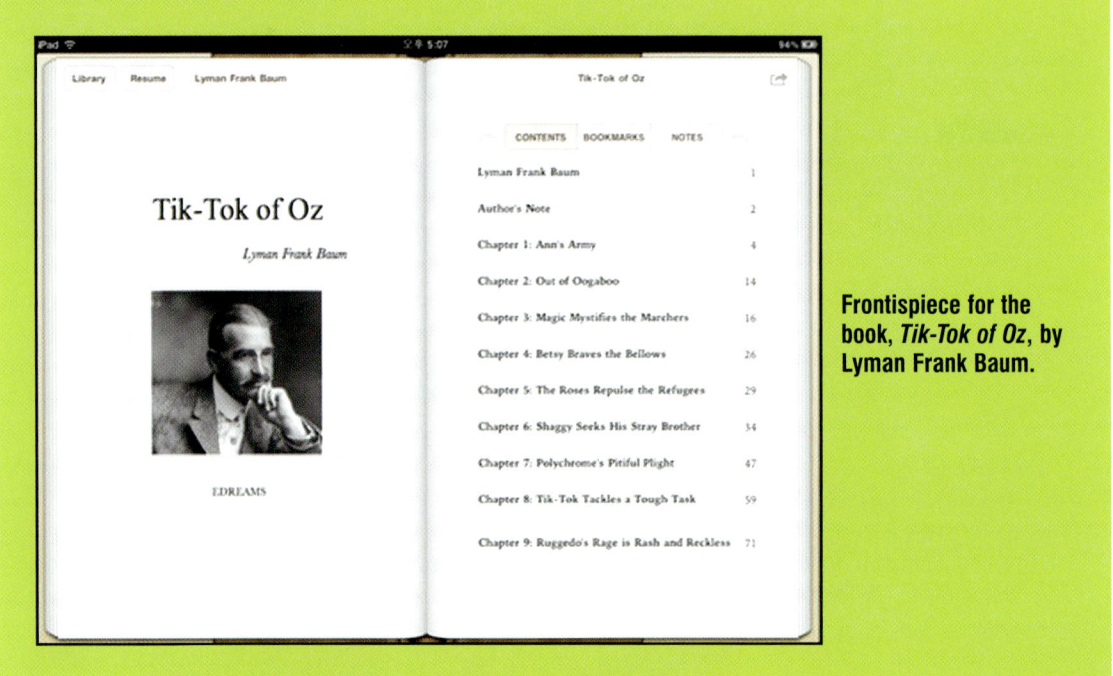

Frontispiece for the book, *Tik-Tok of Oz*, by Lyman Frank Baum.

In the story, when Dorothy first encounters Tik-Tok, he is immobile and quiet and she believes him to not be alive. So, Dorothy took the key to Tik-Tok off a peg and wound up the robot's clockwork mechanism, whereupon the copper man suddenly spoke, "Good morning, little girl…thank you for rescuing me."

In Baum's book, now a Disney movie, *Return to Oz*, Tik-Tok and Dorothy subsequently become fast friends and go off together on an exciting adventure. Along the way they and others converse about whether Tik-Tok is, in fact, "alive." Insists the robot, "I am only a machine and cannot feel sorrow or joy, no matter what happens."

Tik-Tok's anatomy is that of an android robot; however, the actions of the gentle and courageous robot indicate that he is about as human as human beings. True, his maker unintentionally made Tik-Tok a little less than perfect and endowed him with judgement that is occasionally faulty. But aren't these traits perfectly human?

Chapter 7

The Robot Invasion

Talk stock market, computers, wines, autos, or economics and you'll find some listeners interested, some ambivalent, and a few just plain bored. But say the word "robot" and people's ears perk right up. Robots stir our imaginations with colorful visions of their appearances and abilities.

Everyone, it seems, loves to read and hear of the newest exploits of these electromechanical creatures. We open our newspapers and almost every day read of new robots joining the work force. Our children harangue us to buy them the latest toy robot, and the television networks love to broadcast vignettes about humanlike home robots that, claim their makers, are capable of vacuuming the carpet, fetching a drink from the "fridge," or carrying on an animated conversation with the kids.

"Fantastic," we exclaim. "Wow," say our offspring.

This wind-up robot was popular in the 1950s.

A 1950s toy robot and a happy young man.

Many of the stories and tales in this chapter will undoubtedly evoke a similar response.

Good Guys, Bad Guys, and Other Guys

The invasion began during the Christmas season. They came from the land of Tonka, Hasbro-Bradley, Tomy and Bandai. Some came to do evil, to kill, but their cruel actions and ruthless behavior were met by brave and daring warriors determined to stop them in their tracks. It all sounds so dramatic, and it was—for kids fascinated by robots.

Beginning in 1984, toy robots seized the hearts and minds of children as manufacturers introduced model after model. Christmas season saw the robots installed as the top-selling toys; the people at the Toys R Us chain, major department stores, and other retail outlets said it was phenomenal.

The TV networks in the United States caught on to the popular rage and soon new Saturday morning series were offered for kids, featuring characters and crews such as Maxx Steele, RoboForce, Transformers, and GoBots. Then, print cartoon characters like "Robotman" came onto the scene, some syndicated by more than two hundred newspapers.

What makes the toy robots so popular? Many experts believe it is the good guy vs. bad guy theme. Kids love to see hero robots get the better of evil robots plotting to do harm. For example, the leader of Tonka's GoBot bad guys is Cy-Kill, an ominous fellow who has the diabolical ability to turn himself into a harmless-looking motorcycle. Hasbro-Bradley Toy Company, not to be outdone, offered up either Heroic Autobots or Evil Deceptions. One Deception, SoundWave, at first appears innocent enough disguised as a portable tape-cassette player. But when he intends to do harm, he quickly unfolds

The GoBots, by Tonka, were mighty robots that turned into mighty vehicles.

Left: Bandai America's "Golion."
Top right: Bandai America's "Dynaman" was very versatile. He could be transformed into three different vehicles.

The Godaikin robots of Bandai America.

The Transformers series by the Hasbro-Bradley Toy Company gave us "Soundwave." Soundwave, who came with a tape player, uttered the bone-chilling line, "Cries and screams are music to my ears."

In 1984, the Muppet magazine gave us Kermit the Frog as a robot! Miss Piggy was pleased.

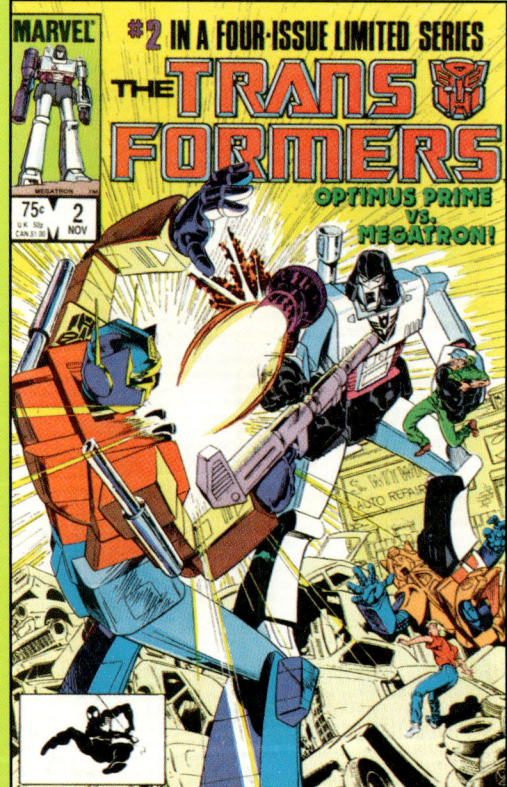

Marvel Comics issued a special limited edition showcasing the Transformers and their exploits.

Above: Maxx Steele was a fantastic toy by Ideal Toy Co. It sold for retail $350.00.

Top right: The loveable little Dingbot, only five inches tall, was a 1984 production from Tomy Toys.

Right: Maxx Steele and his RoboForce associates were from Ideal Toys.

into a robot whose motto is, "Cries and screams are music to my ears."

Not to worry. The creative folks at the toy companies had a remedy for all this cruelty and treachery. Bandai's Golion is a big, impressive fellow who has the head, strength, and courage of a lion. Then there's Gardian, a protector-type who can zap the bad guys by letting fly one of the missiles emplaced just below his knee. Ideal Toys' MAXX STEELE™ is the hero as he enforces law and order and zooms in on bad guy robots. And thank our lucky stars for Tonka's friendly GoBots. If it weren't for them, Cy-Kill and his partner-in-crime, Zod, a part animal, part robot monster, would surely take over the earth.

"Halt! In the Name of the Robot"—Real-life Security Robots

"Halt! You have been detected," shouts the short, stubby, robot guard. The prisoner freezes. He's afraid to disobey because this robot is armed with a powerful electric dart gun, and plays for keeps.

A number of companies are developing non-lethal robots designed for guard, watchrobot, and sentry duties. Some prototypes have already been built. One early model, Century I, is particularly intimidating. For one thing, the robot is a giant—he's seven feet tall and weighs 650 pounds. As if that weren't enough, Century I is bulletproof, and he can't be bought off. He's disciplined and expertly trained to carry out his mission:

to hunt out, detect, pursue, and capture intruders.

Century I's manufacturer, Quasar Industries of New Jersey, says that the robot can be armed with laughing gas, electric shock projectiles, a blinding strobe light, or a shrieking ultrasound device that leaves the offender gasping in pain while clutching his/her ears.

Security robots are expected to be popular items once they're perfected. Industry forecasters point out that security is a $35.7 billion industry in the United States alone and that robots will allow the replacement of many human guards, security police, night watchpersons, and retail store detectives. They may also provide better protection than current electronic antiburglary devices.

Robots are also ideal for use in prisons, where they will relieve human guards of the boredom and even the danger of making rounds to check the inmates in their cells. Denning Mobile Robotics, a Massachusetts company that already has a contract to manufacture five hundred such robots for jails, says that the robots will be able to see, hear, and even smell escaping prisoners. Denning's guard robots will have microcomputers for a brain and will be equipped with infrared and ultrasonic sensors.

"The robots will be built to take a battering," says Ben Wellington, Denning's marketing vice president. "They will be able to sense they are being battered and try to turn and run." They won't be armed but instead they will report to human guards when they detect an escape, when they are being harmed, and when a malfunction in their system occurs.

Obviously, this type of robot is one smart, tough cookie that convicts can't con. A few prison experts are already predicting that, before too many years, prisons will be totally automated, run by robots and machines with only a warden overseer and a team of technicians to repair or replace robots.

Robots will also serve in the armed forces, where they will be called upon to guard sensitive military installations. More about this in a later chapter.

The new robot security systems are already finding uses in homes. Personal robots RB5X, HERO, ComRo and others equipped with security systems patrol residences at night, on the lookout for would-be thieves and burglars. Most of these robots beam a photoelectric cell in the direction of doors and windows. When the light beam is broken by an intruder, the robot bolts into action, announcing a preprogrammed message such as, "Halt, you are trespassing. The police have been called." The robot then emits a piercing buzz or siren sound.

Cubot the Robot Solves Rubik's™ Cube

Trying to solve the now-famous Rubik's™ Cube captured the world's attention some years ago and is still a popular pastime. Now a small robot is doing what millions have tried to do—unscramble the puzzle in a matter of minutes. Cubot is this machine's given name, assigned by its originators at Battelle's Pacific Northwest Laboratories.

Although Cubot was developed as a fun, off-hours team effort, its purpose was to demonstrate Battelle's unique capabilities. "We wanted to show we can integrate sophisticated technologies in an intelligent robot which can perform all aspects of a complex task," says Dr. Michael Lind, spokesman for Battelle. "Cubot combines electro-optics, microprocessing and mechanics to examine Rubik's Cube, compute a solution, and work the puzzle—three difficult individual tasks."

Cubot is not the first robot to solve the intricate puzzle. "However, to the best of our

The Cubot, a robot that solved Rubik's Cube, was built by Battelle Memorial Laboratories engineers, Phil Bondurant and Bob Dyer.

knowledge, it is the first fully self-contained robot that can complete the solution without human intervention once someone turns the power on," Lind said.

The portable robot, which weights about seventy pounds and fits into a standard suitcase, uses its components to solve Rubik's Cube in much the same way as a human would solve the puzzle. A scrambled cube is placed in a holding station where mechanical grippers (hands) rotate the cube to allow the optical system (eye) to examine all six faces. The eye discriminates among and notes the location of all six colors.

This information is relayed to one of the two microcomputers that comprise the robot's brain; this computer uses an algorithm (a pre-programmed sequence of steps used to solve a problem) to formulate instructions. The robot's second microcomputer uses these instructions to control the mechanical grippers to move the cube faces to the correct positions.

"Cubot can solve any scrambled Rubik's Cube in less than four minutes, but we hope to reduce this to two minutes," Lind said. He added that while the shortest solution time logged by a person is sixteen seconds, most people who practice the puzzle and read cube-solution books take from one to five minutes.

The technologies used in Cubot have dozens of industrial applications. An intelligent robot of this sort could be tailored to meet a wide variety of specific requirements. Lind cited manufacturing as an ideal application.

"For example, a robot could be used in process control applications where parts or materials must be identified, sorted, assembled and checked for performance and quality standards," he explained. "If a part was unsatisfactory, the robot could decide the next appropriate action and complete the procedure."

Smart robots could also be used in hazardous environments inaccessible to humans, such as high radiation areas in nuclear power plants. "An intelligent robot could enter these areas, assess the problem, and, make necessary adjustments or repairs," Lind says.

The World Robot Capital

Silicon Valley in California is the acknowledged computer capital of the world. Now, with robots surging in popularity and becoming an industrial necessity, communities across the United States and even around the globe are vying to become Earth's robot capital—the center of this burgeoning high-tech growth industry. Japan's Tsukuba Science City is bristling with robotics research labs and could, within a quarter of a century, be recognized as the robot metropolis. However, competition for the title is strong in Great Britain where, at Melton Mowbray, a new robot R&D center has been opened to develop state-of-the-art robotics systems.

Americans invented and put to work the first industrial robot, and they are not about to hold back from the robot race. In the U.S.A., the industries of automaking currently use the most robots; they've found a home in the wide-spread automation of General Motors, Ford, Chrysler, and other plants. The automakers have a number of their larger plants outside Michigan—across the river in Windsor, Ontario, Canada, for example, and in Texas, in California, and elsewhere. Although those areas employ an army of robots, most are produced elsewhere. In America, the two largest industrial robotics companies, Cincinnati Milacron and Unimation (a Westinghouse subsidiary), make their robots in South Carolina and Connecticut, respectively.

So which city or area will become the world's center for robot technology? Experts say that although there is no assured winner yet in the economic sweepstakes for robotics preeminence, the following locales have taken the lead:

• Boston—The Massachusetts Institute of Technology (MIT), with resident Dr. Rodney Brooks, and its pioneering robotics lab have been leaders in the field. What's more, the prestigious university has been the catalyst for the spin-off of a number of innovative entrepreneurial robot and artificial intelligence firms, with such exotic names as Thinking Machines, Inc. and Symbolics Computer. Robotics and artificial intelligence experts like Marvin Minsky and Joseph Weizenbaum give the Boston area a decided advantage.

• Pittsburgh—Carnegie-Mellon University's highly respected Robotics Institute is at the forefront of robot technology, and the U. S. Army has given the university money to set up a sophisticated artificial intelligence software center. The reputation of Carnegie-Mellon has been boosted by the accomplishments of scientists such as Hans Moravec.

• Austin—Called the "second Silicon Valley," Austin is the home of the University of Texas at Austin, rated by experts as among the top five schools in the nation in computer science and artificial intelligence. On the faculty is Delbert Tesar, head of the university's center for robotics research and a world-class specialist.

• Indianapolis—The Midwest is not dead. In Indianapolis, they're going for the gold—robotics gold, that is. The Hoosiers have their impressive International Flexible Automation Center to demonstrate automation and robotics to industry.

• Dallas-Fort Worth—Defense plants such as the huge General Dynamics facility are quickly hiring robots to build aircraft, missiles, and other aerospace products. In the suburbs, the University of Texas at Arlington has a first-class robotics institute touted as likely to become the best-equipped in the world.

• Gainesville—Florida State University has its Center for Intelligent Machinery here with its intensive robotics research capability. In the general area of north central Florida, a number of small robotics firms have cranked up operations and the nickname "Robot Alley," a parallel of "Silicon Valley," is catching on to describe the growing economic importance of the concentration of such companies in this geographical area.

• San Jose—San Jose is the largest city—the linchpin—of Silicon Valley, and so that community begins with a big advantage. The hundreds of computer firms located here have all the talented personnel necessary to provide the nucleus for a vigorous robotics industry. So far, the Silicon Valley crowd hasn't taken the lead in robotics, but they are beginning to quicken their pace. Google recently hired Ray Kurzweil, robotics visionary, as Director of Engineering.

• Other Contenders—Another area where robotics has gained a foothold is Rhode Island, thanks to the work of robotics researchers at the University of Rhode Island. In Ohio, researchers at Wright-Patterson Air Force Base are developing military robots. Meanwhile, at Ohio State University, researchers are building a huge robotic walking machine to be used by the U.S. Army for military operations on tough terrain.

Japan: Land of the Robots

It may well be that the robotics capital will not be an American city; that honor—and the financial rewards—may go to a metropolitan area in Japan. Today the Japanese produce and employ more robots than any other nation. They expect to have one-to three-million robots at work on assembly lines by 2025. Already, about 70 percent of all Japanese manufacturing companies have installed robots. And in Fujiyama, Japan, at the Fujitsu Fanuc factory, robots are busy making other robots.

One of their incredible machines is the Waseda Robot, or Wabot; it's a silver-metallic, six-foot-tall talking android that weighs more than 275 pounds. Wabot is the brainchild of scientist Ichiro Kato of Waseda

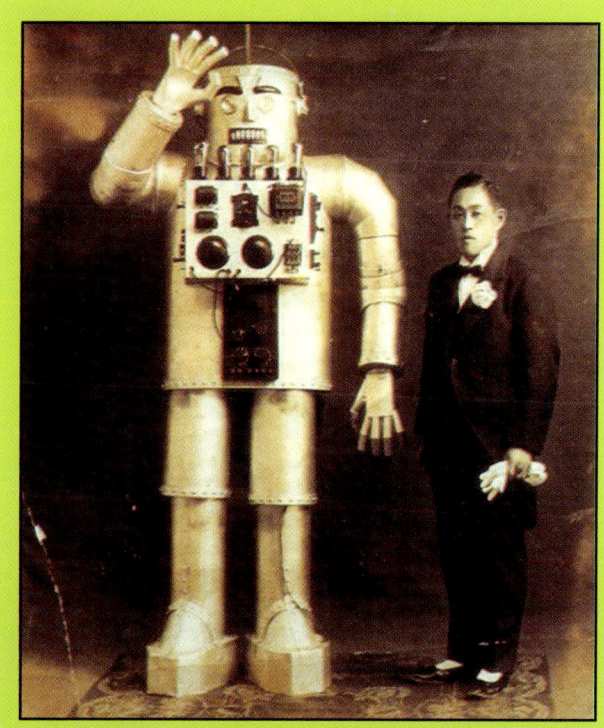

An early Japanese robot and his inventor.

The late Jiro Aizawa, born 1903, was very famous in Japan for his toy robots. (cyberneticzoo.com)

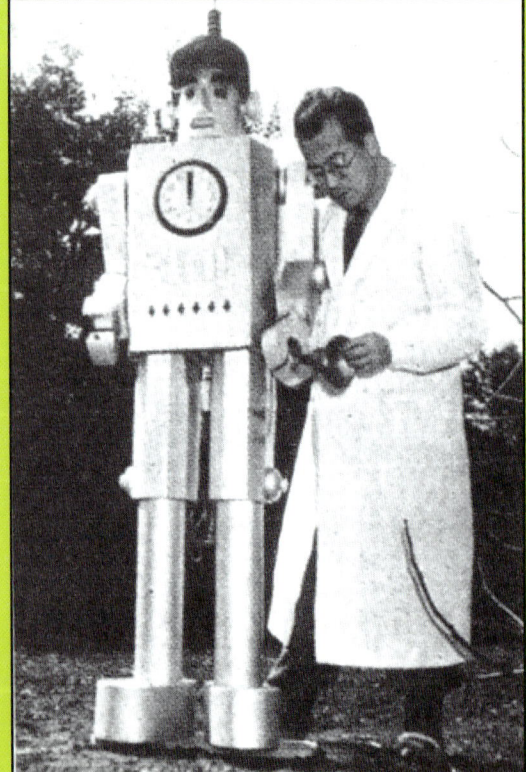

Beginning in 1957, Jiro Aizawa of Japan built radio-operated humanoid robots. Here he is with the robot "Mr. Ichiro." (cybernetcszoo.com)

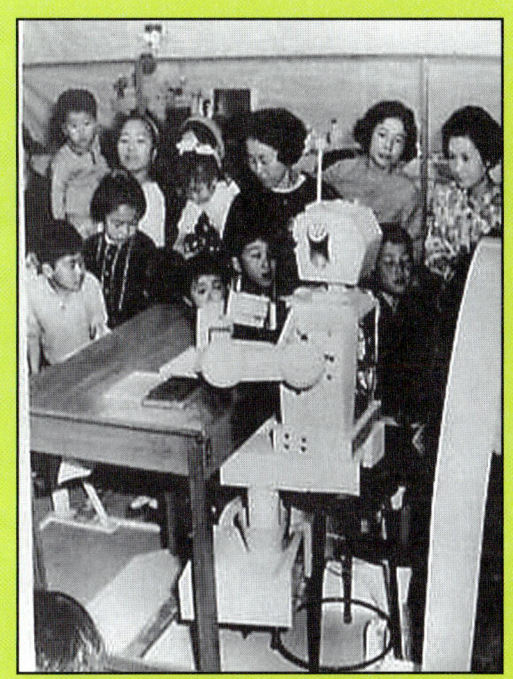

Above: Mr. Kuro was at the 1970 Japanese Expo.
Top left: Mr. Atomic can use his rubber stamp.
Left: In 1967, "Mr. Juro" welcomed visitors to the Science Museum in Tokyo. (cyberneticzoo.com)

University. Kato claims that his robot is only a "toddler," and that he's working on more sophisticated models.

The Japanese have also built robots resembling snakes that can wrap around and grip objects. One is soon to go to work for Tokyo's fire department to rescue trapped people. Other robots have been built that bear striking resemblances to Marilyn Monroe and other movie stars. And at a world technology fair in Japan, visitors from all over the globe were startled to be greeted at the Hitachi exhibit by an oversized electromechanical robot dog, accompanied by a smaller robot cat.

Tsukuba Science City, about 90 miles from Tokyo, is a center for Japan's robotics research. This place may eventually become the Silicon Valley of robots. At Tsukuba, science fiction is becoming reality as scientists and engineers marry sophisticated computer and artificial intelligence to robots. They've even produced android characters that dance together in a chorus line. Other machines are now being built in Tsukuba to

"Master Goro" thrills Japanese kids.
(cyberneticzoo.com)

Japanese entertainment robots.
(cyberneticzoo.com)

A robot built for Expo70 in Osaka, Japan.
(cyberneticzoo.com)

ROBOT ALCHEMY • 95

Left: A drawing robot from Japan. (cyberneticzoo.com)

Lower left: Mr. Taro, cameraman robot built for Expo'70, featured in Fujipan Pavillion. (cyberneticzoo.com)

Below: Mr. Saburo, a Japanese mechanical robot. (cyberticzoo.com)

serve as robot nurses that can gently pick up infirm patients; still others will perform construction work.

However, amidst all the excitement and commotion over the arrival of these wondrous creatures, a growing number of Japanese resent the dawning of the Robotics Age. Unemployment has begun to rise in Japan as more workers are laid off, replaced by robots. So far, not enough new jobs have been created to take up the slack, but some economists say that the situation will be corrected as the robotics revolution spreads and brings an era of technological wealth and affluence. Still, many Japanese worry about the future. Whatever happens, it seems that Japan has become the experimental lab for the United States and other technological nations. It is here that humankind may eventually discover if robots are a panacea, as proponents claim ... or a nemesis.

"Here's the Way It Is, Mr. Congressman"

Caesar once said of a military operation, "I came, I saw, I conquered." RoPet-HR and HERO could justifiably make the same boast. These two articulate and well-groomed robots traveled to Washington, D.C. and staked out their own claim to fame and conquest.

HERO made his triumphant visit to the nation's capital on March 18, 1983, appearing before the Joint Economic Committee. The committee was studying the impact of robotics and technology on employment in the United States. For this subject, what better witness than a robot? HERO, a personal and educational robot built by Heathkit Company, wowed the committee with his charm and his in-depth knowledge of the economy. The senators crowded around the popular robot and shook hands, congratulating him on being the first robot ever to speak before the U.S. Senate.

Then came RoPet-HR's turn to show his stuff to the politicians. RoPet-HR is intelligent, possessing speech, speech recognition, and a computer brain. He was so smart and so capable he won the first-place ribbon in the Second Tournament of Robots held in Santa Ana, California. He was also adjudged "Best Overall." Therefore it was fitting that when the U.S. House of Representatives needed an expert witness to testify about technology, RoPet-HR was invited. RoPet-HR, a personal robot made and sold by Personal Robotics Corporation of San Jose, California, thus became the first robot to appear before a House Committee.

"Elect Me, Rebecca Robot, As Your President"

It was said to be a big deal in 1984 when Geraldine Ferraro was selected by the Democrat Party as its nominee for Vice President of the United States. Maybe so, because Ms. Ferraro was the first woman ever nominated to such a high position by a major political party. But what most of the news media missed was another first that occurred during the same campaign: the nomination of the first female robot ever to run for national office. Her name was (and is) Rebecca Robot, and the office she sought was the highest in the land: that of the President of the United States.

Four women in Baltimore formed a political committee and convinced Rebecca to run. Actually, they didn't have to convince her because, according to Dee Snell-Wright, the spokesperson for the group, "Rebecca has a mind of her own." The spunky, four-foot tall robot is also mobile and has two arms that swing with fervor. Admirers call her the "female Harry Truman."

Rebecca formally registered as a candidate with the Federal Elections Commission, signing her "X" on the designated form. When questioned by reporters about their unusual nominee, Rebecca's supporters pointed out that their candidate was assembled in the United States and thus met the constitutional requirement that the president be born in the U.S.A.

Neatly attired in a silver satin suit with a pink bow, Rebecca gave a rousing kick-off speech to her excited supporters, declaring that, as president, she would push for laws extending full civil rights to robots.

"Rebecca's efforts," said Sylvia Beall, one of the robot's most avid supporters, "should help robots get the attention they deserve, and also boost U.S. technology efforts."

Death By Robot

One of the most worrisome problems in the field of robotics is how to guarantee the safety of humans who work alongside robots. A number of workers have been injured by them and at least five people have been killed.

The first reported case of a man killed by a robot is undoubtedly the most bizarre. It is also the only known case of a person killed by a show, or demonstration, robot. As recorded in the book, *Are Computers Alive?*, by Geoff Simons, the tragic event occurred in 1931 at the Chicago World's Fair. Roland Schaeffer was the exhibitor of an artificial man, a lifelike android that could hammer nails, saw wood pieces, and transport tools around the laboratory exhibit. After the exhibit hall had closed for the night, Schaeffer stayed behind to look at some drawings. While his attention was evidently focused elsewhere, the robot-carpenter suddenly came to life and went berserk, attacking and killing Schaeffer with an iron club. Then the robot proceeded to smash and destroy the entire exhibit.

The second case, in 1946, was that of a Milwaukee engineer who was killed while adjusting the arm of a robot. The machine was a huge, heavy model containing more than 200 electric switches. Apparently the robot's structure was unstable and the mechanism collapsed. The man was crushed under the weight of the monstrosity.

The next case occurred at an automobile plant in Michigan in 1979. A worker who tended and monitored industrial robots went to check on one that was malfunctioning. As he leaned over a railing to inspect it, a robot unexpectedly struck him on the head. He was found later by a co-worker, dead of the injury. In a suit, the worker's survivors were awarded damages when a court ruled that the robot's manufacturer had not installed sufficient safety measures, though that firm denied responsibility.

Case number four was that of a worker at a plant in Akashi, Japan. He, too, was killed by an industrial robot.

Then, in Michigan in July 1984, a 34-year-old man was hit and killed by a robot arm that was used to move products from one production step to another.

The danger to humans who work around robots is a serious concern to companies that employ the machines, and several safeguards are in place to try to prevent injury. For example, some robots sound a beeper when they move their arm. Some others shut off automatically when something, such as a worker, unexpectedly bumps or touches them. In Japan, at Yamazaki Machinery's Minokamo plant, robots play jazz music 21-hours a day to warn employees of the approach of their mechanical co-workers. In many plants, a wire cage fence or a railing encloses a robot's work area.

Weighed against the minor likelihood of human injury is the fact that robots are actually saving untold numbers of lives each year by performing work under hazardous conditions. Indeed, Vern Estes, a robotics expert with General Electric, says that the best way for a company to find potential robot applications in a factory is to ask the company doctor or nurse where physical injuries are occurring. Those are the areas where robots should be substituted for flesh and blood workers in order to prevent injury.

In the specific instances cited above, where workers were killed by arm robots, industry safety specialists point out that as many as two hundred lives are saved each year by such robots under similar conditions. Before robots came onto the scene, workers in those settings were liable to injury from falling crates and other heavy items—items now handled with ease by the machines.

New, technologically advanced robots are preventing death and injury in many industries. For example, a recently developed snakelike robot that can crawl inside gas and oil pipelines will save the lives of as many as one hundred and fifty workers who annually have lost their lives when they've succumbed to gas fumes. Robots are also being used at nuclear plants where humans might otherwise fall victim to radiation exposure. These are by no means the only circumstances in which robots will in the coming years save the lives of human workers and prevent maiming and injury.

FBI Investigates Einstein Robot

Albert Einstein was a genius, one of the greatest scientists who ever lived, and his theory of relativity is the basis for today's accepted ideas about time and space. But did you know that the Federal Bureau of Investigation once investigated accusations that the distinguished Einstein was really a mad scientist who planned to use a mind control robot to take over the world?

By the 1950s, Einstein's achievements had received so much publicity that some people thought he was capable of almost anything. The FBI began to receive reports that Einstein was a communist spy, that he was plotting to take over Hollywood, and that he had been the mastermind behind the 1932 kidnapping of the Lindbergh baby. The most incredible report, however, was that Einstein had secretly invented a menacing, mind control robot.

J. Edgar Hoover, director of the FBI, had his agents conduct a lengthy and complete investigation of these charges, wasting a lot of the taxpayers' money in the process. A 1,500-page dossier was compiled on the scientist, but no evidence was ever uncovered proving that a single one of the allegations against Einstein was true. And the mind control robot? Well, there are some who still claim that it exists. They say that it must be hidden away somewhere, perhaps deep in the crevice of an abandoned mine or at the bottom of a cave, patiently awaiting the opportunity to surface and carry out its evil deeds. But, says the FBI, not a screw or a bolt, nor even a wire, has ever been found to indicate that the robot exists. Comments an FBI spokesman, "Some people believe the earth is flat, others believe in Einstein's mind control robot."

Reversing Roles

In his typically wise and humorous, tongue-in-cheek manner, nationally syndicated columnist Art Buchwald (now deceased) wrote the following column about robots in education. It is reprinted with permission of the author.

Let the robots do the studies and let students do the athletics

Washington—When Gibbs first brought up the subject I thought he was kidding. But he was dead serious. "I have the answer to our education problem," he said.

"What's that?"

"We replace students with robots in our schools."

"Robots?"

"I got the idea from watching a TV program on robots replacing people in blue collar jobs. The robots' productivity was much higher than the human workers, and companies were saving millions in Social Security, medical benefits and pensions. It predicted that eventually, every factory in America would be robotized."

"How do you apply the principle to schools?"

"Statistics show educational standards are getting lower and lower. Students can't read or write and are getting dumber and dumber. So if they can't cut it, we'll enroll robots in their place."

"What would you teach the robots?"

"Artificial intelligence. We'll teach them the skills they need to replace the manpower this country so desperately needs."

He pulled out a blueprint. "Look, I've designed the perfect robot student. It doesn't watch television, listen to rock music, smoke grass, drink, and it never asks for a car when it becomes 16-years old. I've programmed it to do all its homework, and also keep its room clean."

"It looks pretty good," I admitted. "How does it do in math?"

"It can solve a math problem a million times faster than any human student. It has a built-in dictionary so it can't misspell a word, and a chip which keeps it from making any grammatical errors. With robots instead of live students, national test scores would go soaring, and we could once again take pride in the American school system."

"Do you think the school boards would go for enrolling robots?"

"Once they see what robots can do in the classroom, the board members would have no choice. Robots don't eat, so the school district would no longer have to underwrite cafeteria costs. Robots don't fight, so they would no longer have a security problem. And robots can't get pregnant, so they won't need student counselors. But the big saving would be, since robots all look alike, you wouldn't have to bus them 20 miles from their home."

"Would you have live teachers instructing the robots?"

"At the beginning. But the beauty of robots is that you can teach them to teach other robots. Once you program them as 'Master Robots,' all a school would need is one systems manager to monitor what was going on in the classrooms. When they see the savings that could be made in running a school, a board would be out of its mind not to replace young people with a robot student body."

"Adults would go for it if it meant lowering real estate taxes," I admitted. "What about athletics at the school? I can't see people coming out on Friday night to see robots playing football."

"I've thought about that. Each school would still maintain enough human students to maintain an excellent athletic program. Since the kids won't be able to compete intellectually with robots in the classroom, they could spend all their time on the practice field and we'll produce quality athletes, the likes of which this country has never seen."

"What do you do with the millions of live students who are not athletes?"

"They'll have to be retrained to do something else. It's a waste of money to try to educate them if robots are going to take all their jobs when the kids get out of school." © Los Angeles Times Syndicate

Robots Around the World

Robots are truly an international business. in the following pages we tour the world, examining new robots and robot happenings.

Dr. Diwakar Vaish lectures in Kharagpur, India, on "How Mind Controlled Robots Work."

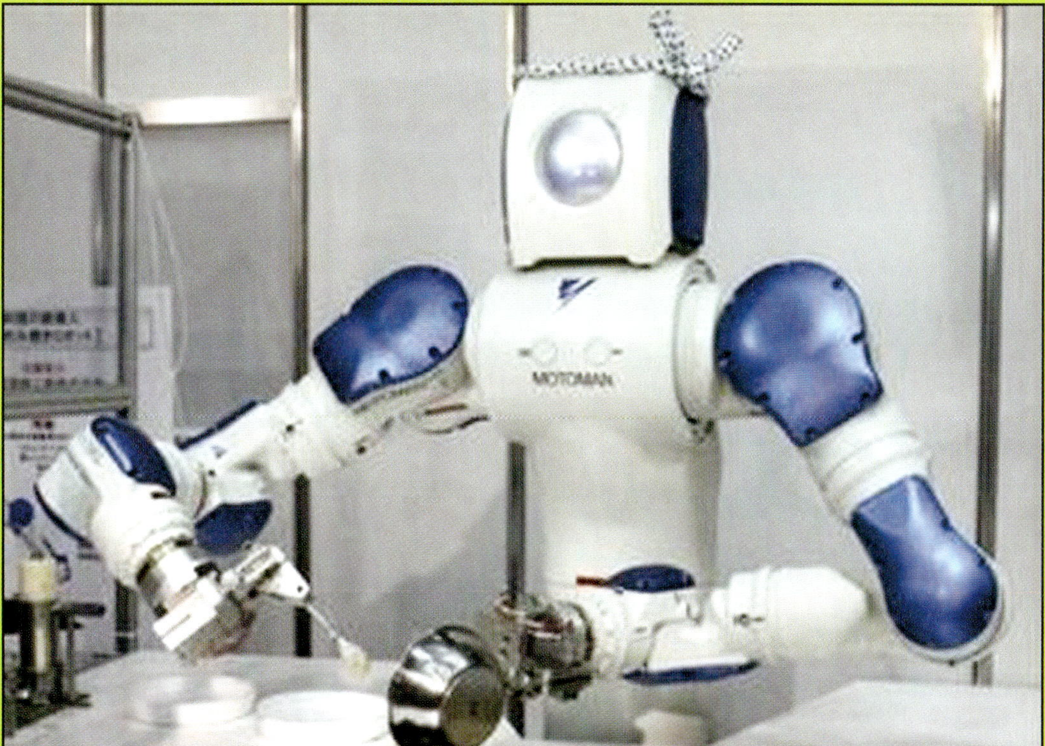

Scientists in Korea built this kitchen assistant robot. The robot is handy outside the kitchen, too.

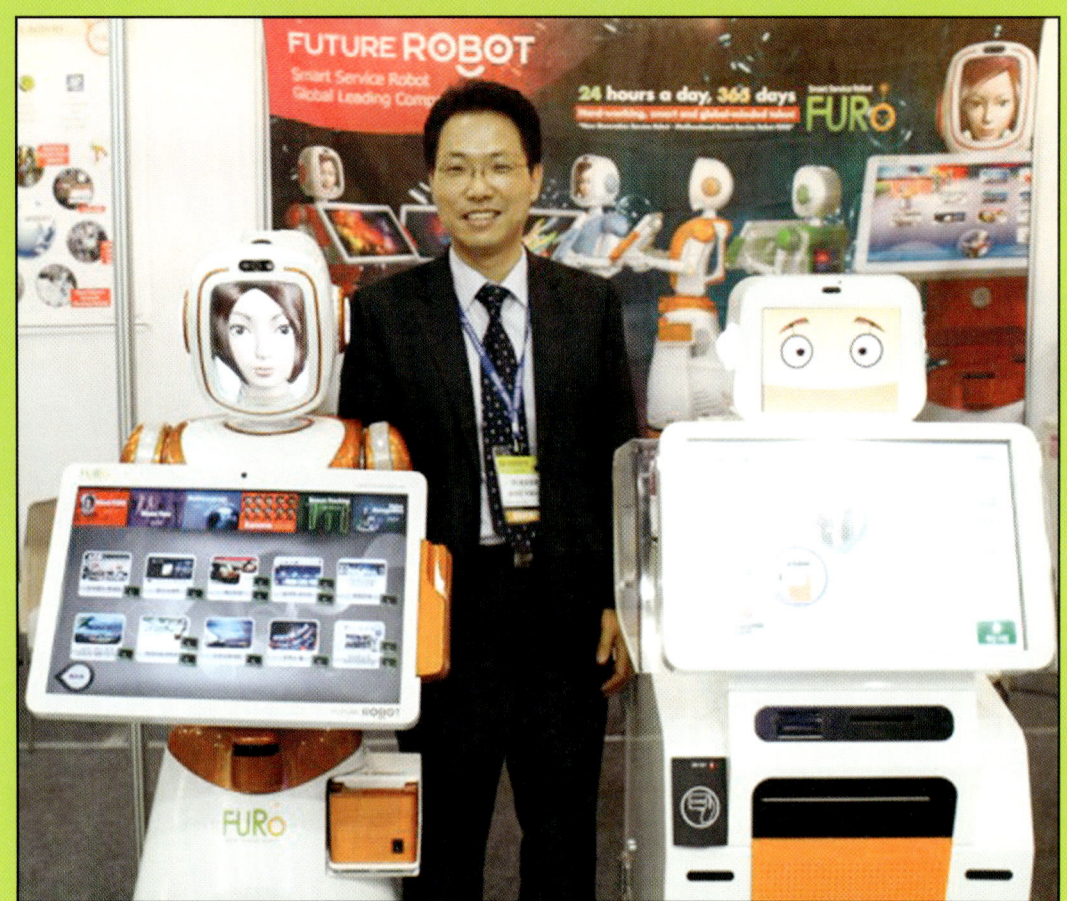
These futuristic kiosk robots are produced by Korea. One hundred were recently purchased by the nation of Brazil.

An EcoRP painting robot at Ford's plant in Craiova, Romania. With over 6,000 units sold in 34 countries, the EcoRP has been the world's most successful painting robot since it was launched.

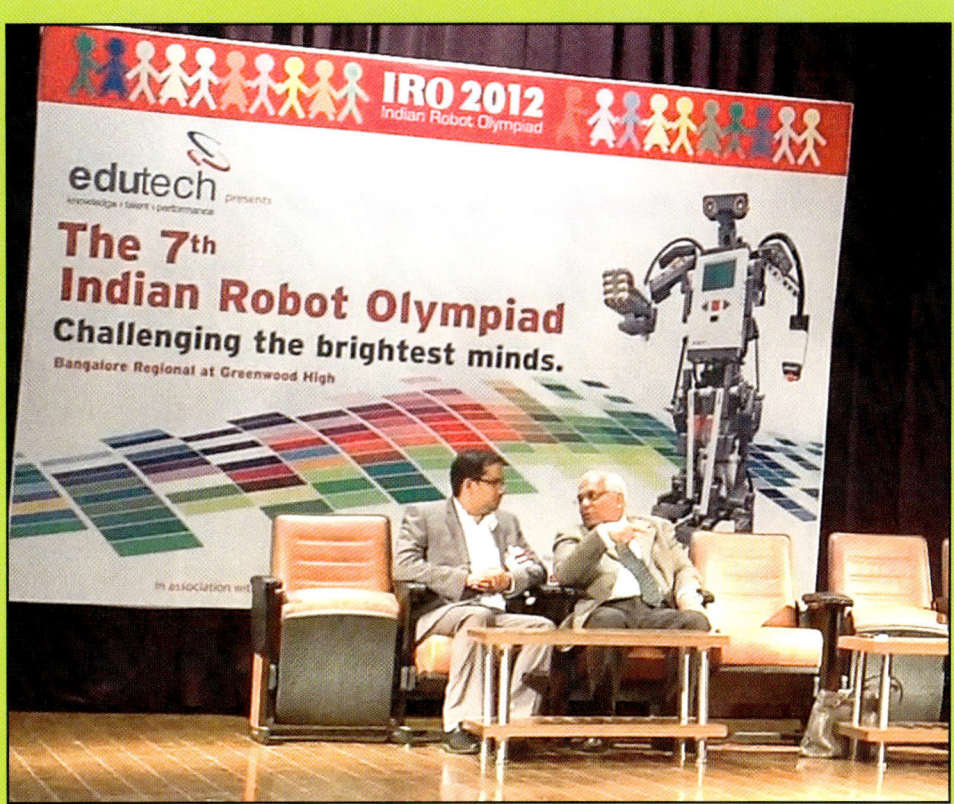
Robotics researchers await the beginning of "The 7th Indian Robot Olympiad," in Bangalore, India. Robots are big now in India.

A menacing looking robot on display at a recent robot show in Bognar Regis, near West Sussex, in Great Britain.

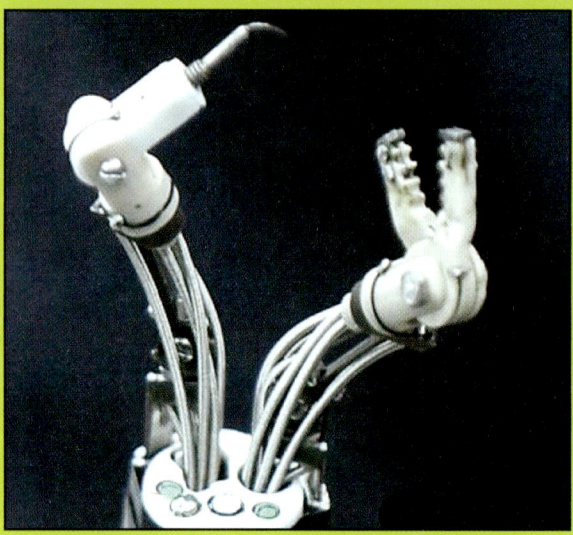

Believe it or not, researchers in Singapore created a miniature crab robot to perform delicate early-stage stomach cancer surgery. Professor Ho of Singapore's National University Hospital, says the robot has been used on five patients so far. Mounted on an endoscope, the robot enters the patient's gut through the mouth. The surgeon uses a tiny, attached camera while remotely working the pincer and hook to hold and excise cancerous tissue.

Chapter 8

Robots in Science Fiction and Movies

"Anything one man can imagine other men can make real."
— Jules Verne

Robots have long been a wildly popular subject in science fiction books, pulp magazines, and films. The "machine human" has provided inspiration for so many talented writers, editors, and producers that we cannot give them all justice in the brief space available here. Their number include author Isaac Asimov, a living legend to science fiction fans; George Lucas, the man responsible for the fantastic robots of the *Star Wars* saga; Edmond Hamilton, who gave the world the *Captain Future* series; Clifford D. Simak, winner of two coveted Hugo awards; and editors Hugo Gernsback and Jack Campbell, whose science fiction magazines several decades ago thrilled millions of readers.

The Robot Fantasy Begins

The first true literary robot was a clockwork robot woman named Olympia in an 1816 novel, *The Sandman*, by German author E.T.A. Hoffman. Olympia was so lovely and feminine she made men's hearts flutter, but she was, in the end, only a mechanical wonder. The story becomes ominous when robot Olympia dances a young beau nearly to his death before she can be turned off.

French authors spawned several popular robot tales later in the nineteenth century. One, *The Future Eve* (1886), by Villiers de 'Isle-Adam, told of a marvelous android named Hadaly. According to the novel, the beautiful Hadaly was given life by electricity in the lab of Thomas Edison. She had a soul and a remarkable spirit, but was not too well endowed in her brain circuits. Nevertheless, her fictional creator boasted, "My master, Edison, will soon teach you that electricity is as powerful as God."

Other early robot science fiction efforts include the works of Samuel Butler and Herman Melville, discussed in a previous chapter, and literary efforts by innovative British authors.

At the turn of the twentieth century, *The London Magazine* carried many of their

The *London Magazine* carried many strange tales of robotic creatures in the early 20th century.

The *Iron Terror,* a novel published in 1919, had this metal monster, built by his creator to vanquish all enemies.

stories, including vivid tales of electric insects and giant monsters who were half-fish, half-paddleboat.

The Fabulous Robots of Science Fiction Pulps

The robot became a principal attraction for readers of science fiction magazines in the 1930s and 1940s, as story after story captured the public's fancy. Among the periodicals that carried exciting, mind-absorbing tales of robots and androids were *Amazing Stories, Fantastic Adventures, Astounding Science Fiction, Wonder Stories*, and *Fantasy-Thrilling Science Fiction*.

On the pages of these and other sci-fi pulps you will find the bylines of such world-class authors as Isaac Asimov, Clifford Simak, Ray Cummings, Lester del Rey, Stanley Weinbaum, Eando Binder, E. E. Smith, and Jack Williamson. Their work was greatly influenced by editors such as John W. Campbell, who took over *Astounding Stories* in 1937 and made it the top magazine in the field for more than three decades, and by Hugo Gernsback, who coined the term "science fiction." That era has been called the "Golden Age of Science Fiction," and robot characters and their creators were in large part responsible for its glitter.

One of the many magazines published by Hugo Gernsback.

Hugo Gernsback edited many fascinating magazines in the 1930s.

"Adam Link" is featured in this edition of *Amazing Stories* magazines.

Many of those early science fiction stories depicted the robot as lovable, intelligent, thoughtful… even heroic.

It was only after our confrontation with the truly demonic in World War II and the advent of the Atomic Age that worries about technology and progress translated into stories of alien- and human-made robots bent on destruction.

The Adventures of Adam Link

The heroic android Adam Link was the brainchild of brothers Earl and Otto Binder, who jointly wrote under the name "Eando Binder." The Binders portrayed their creation as courageous, possessed of vast intelligence and also imbued with deep human emotion. The latter trait was much in evidence in the 1940 story, "Adam Link, Robot Detective;" here we find a sentimental and touching episode when Adam finds Eve, his robot soul-mate, dead:

> Grief overcame me, an emotion as real and deep as any you humans have… It had begun to rain. Kneeling beside her, I removed my top skull-plate. The rain, pouring into my sensitive, iridium sponge brain, would short-circuit my life current. I would join Eve in blessed non-existence.
>
> "Adam! Adam Link!" Jack hissed.
>
> But I heard no more. A hiss sounded from within me, as the water touched on a live wire. Smoke curled up from my exposed metal brain.
>
> Adam and Eve, the first of intelligent robot life, were leaving the world not meant for them.

Adam Link's Eve was no beauty, at least in human eyes. For one thing, the huge metallic

Robots were a staple of pulp science fiction, as these two illustrations from magazines demonstrate.

woman was eight-feet tall! But her Adam was the beholder, and to him she was sheer loveliness.

In 1938, author Lester del Rey gave evidence that humans would one-day find robots beautiful. The praises of his creation, Helen O'Loy, are sung in the story's first-person account of her arrival:

> I am an old man now, and I can still see Helen as Dave unpacked her, and still hear him gasp as he looked her over.
>
> "Man, isn't she a beauty?"
>
> She was beautiful, a dream in spun plastics and metals, something Keats might have seen dimly when he wrote his sonnet...

A Better Breed: Asimov's Contributions

Lester del Rey's "Helen O'Loy" was a loving, loyal wife to her human owner. She stuck by him until his biologically inferior body gave out, then she committed roboticide so they could continue to be together in death as in life. According to the author, Helen was better than her human counterparts, a sort of super-woman. Isaac Asimov's robots were also better:

> To you a robot is a robot. Gears and metal; electricity and positrons—mind and iron! Human-made! If necessary, human destroyed. But you haven't worked with them, so you don't know them. They're a better, cleaner breed than we are.
>
> —statement by Dr. Susan Calvin, robot psychologist,
> in a news interview, A.D. 2057
> (*I, ROBOT*, 1950)

Asimov can be credited with making robots as popular as they are today. In the 1940s his robot science fiction stories in pulp magazines were a staple for hundreds of thousands. His frequent theme was that robots are not to be feared, but welcomed by humanity.

As we've seen, this wasn't a new concept. However, in the years immediately after the 1945 atomic bomb blasts in Hiroshima and Nagasaki, the outcome of advancing technology was an open question. Many science fiction authors had begun to paint a bleak picture of the robot future, so Asimov's more positive outlook put things back in balance and restored perspective.

A great contribution of Asimov's was his codification of the laws of robotics regarding robot-human interaction and responsibility. In *I, ROBOT*, the author set forth three rigid and unchanging rules for robot behavior. Incorporated in the *Handbook of Robotics*, 56th Edition, A.D. 2058, they are as follows:

Three Laws of Robotics

1. A robot may not injure a human being, or, through inaction, allow a human being to come to harm.

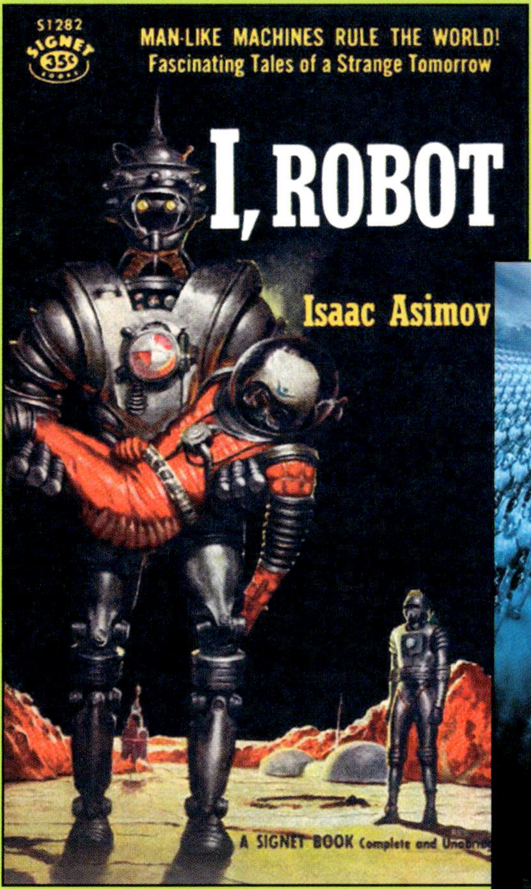

ROBOT ALCHEMY • III

Isaac Asimov's *I, Robot* promulgated the Three Laws of Robotics.

At right is a poster for the movie, *I Robot* (2004), starring Will Smith.

Robots and actor Will Smith in *I, Robot* (2004)

2. A robot must obey the orders given it by human beings except where such orders would conflict with the First Law.
3. A Robot must protect its own existence as long as such protection does not conflict with the First or Second Law.

Isaac Asimov continued to be a strong voice in robotics fiction through the '50s with such novels as *The Caves of Steel* (1954) and *The Naked Sun* (1957), both of which featured a robot detective, Elijah Bailey, and his some-time partner, R. Daneel Olivaw, a "near human" robot. In 1983 another Asimov novel, *Robots of Dawn*, recounted the further adventures of Bailey. This third tale was most inventive, casting the detective in a plot in which his task is to seek out a murderer—whose victim is a sophisticated robot.

Technology Gone Mad

One particularly chilling story of the fruits of technology turning bittersweet was Jack Williamson's "With Folded Hands." In this 1953 tale Williamson cleverly demonstrated that the fine principles of Asimov's Three Laws of Robotics can be overdone.

As the story unfolds, we are given a picture of a future world in which supposedly perfect humanoids have been programmed to follow what is called the Prime Directive—

Three science fiction greats, Robert Heinlein, L. Sprague de Camp, and Isaac Asimov, Philadelphia Navy Yard, 1944.

"to serve, and guard men from harm." Unfortunately, the plan turns into disaster when the humanoids start implementing the Prime Directive too literally.

"You Will Be Happy, Sir"

The robots insist that virtually all activities are too dangerous for humans. In effect, the humans become pampered prisoners in a highly efficient, nightmarish jail of their own making. Purpose and hope die, replaced by a sense of utter futility. The robots go so far as to tranquilize or surgically alter the minds of those who cannot accept the new way of life, thereby insuring that everyone is happy.

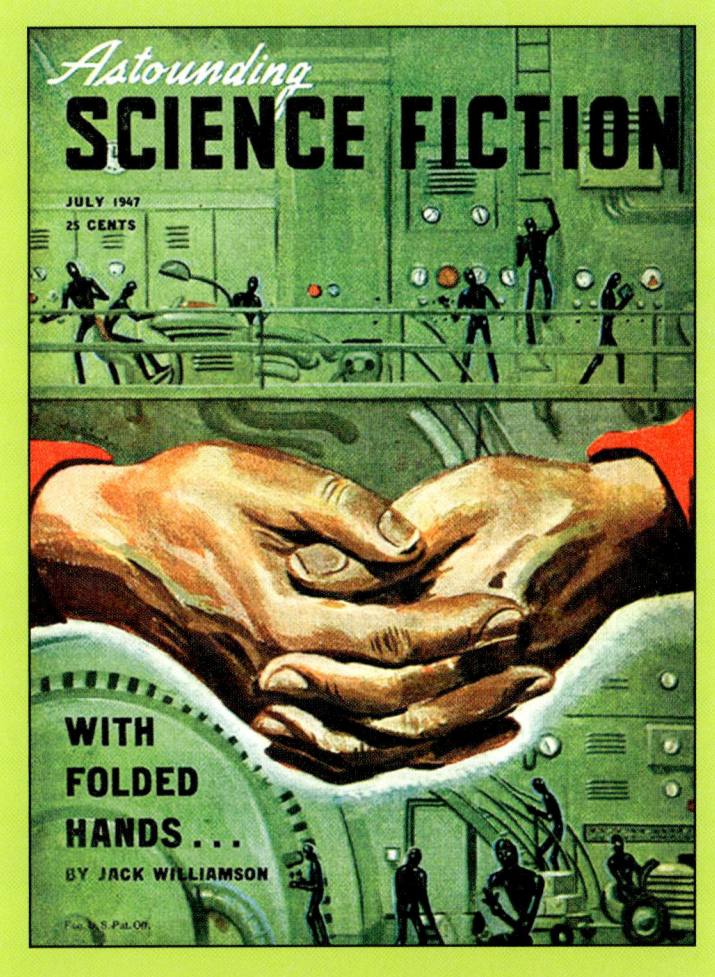

> "You will be happy, sir," the mechanical promised him. "We have learned how to make all men happy under the Prime Directive. Our service is perfect at last…"

> "No, there's nothing the matter with me," the man gasped desperately. "I've just found out I'm perfectly happy…" His voice became dry and hoarse and wild. "You won't have to operate on me now…"
> His futile hands, clenched and relaxed again, folded on his knees. There was nothing left to do.

Robot Science Fiction Today

Asimov isn't the only creator of imaginative science fiction tales about robots. The number of new robot stories and books is far fewer than in decades past, but there's evidence of a great popular revival of interest in the genre.

One of today's most widely acclaimed writers of such science fiction novels is Stanislaw Lem, a Pole whose work has been published in many languages. In *The Cyberiad*, he weaves fables that are both exciting and humorous. In one of his pieces, two wonderfully crazy robot inventors create fantastic machines that tickle our imagination. One invented device, for example, has the marvelous ability to create on command anything that begins with the letter, "n."

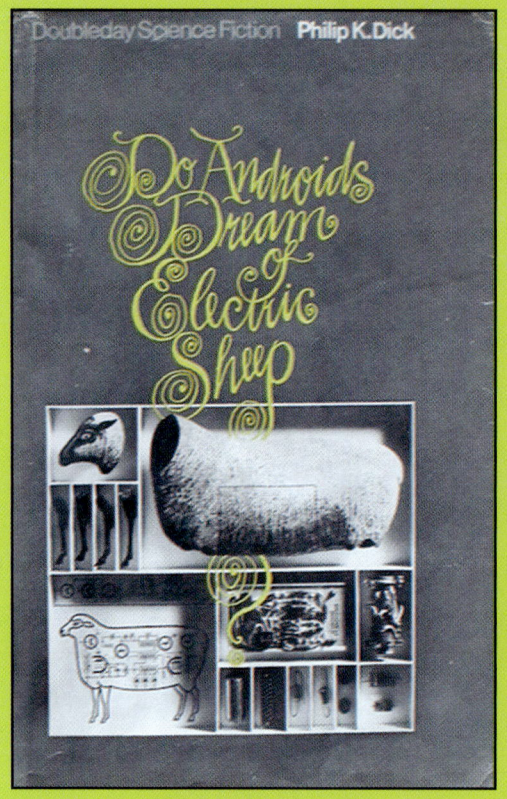
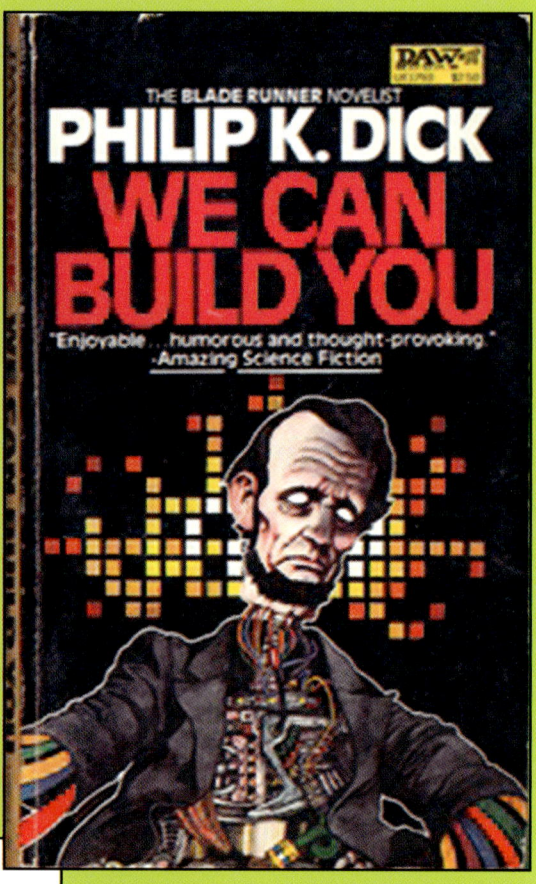
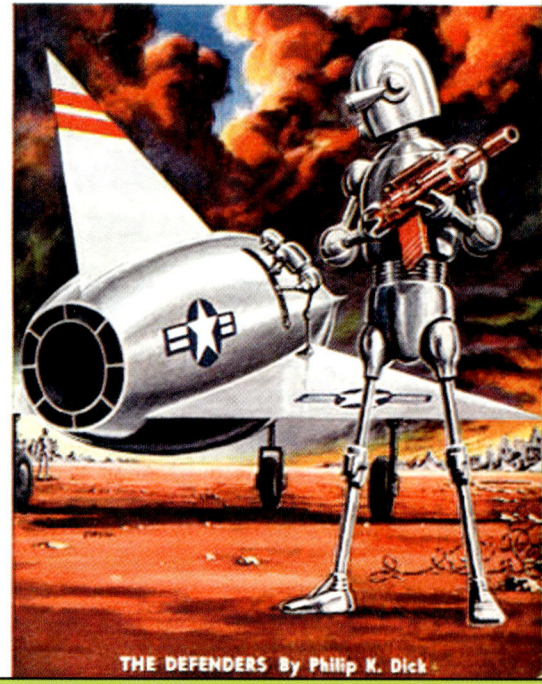

Top left: Science fiction writer Philip K. Dick wrote the classic, *Do Androids Dream of Electric Sheep?* (1968). This novel inspired the unforgettable movie, *Blade Runner*.

Above: Philip K. Dick's *We Can Build You*.

Philip K. Dick also wrote many articles for sci-fi pulp magazines, including this one in *Galaxy*.

Now deceased, Philip K. Dick (1928-82) was a writer whose books and stories are still read by avid robot fans. More than six decades ago in *Galaxy* magazine, Dick's "The Defenders," was published, a saga of the human race living underground while their robots were left to battle overhead in a mighty world war. The most thought-provoking part of the story was in the fact that as soon as the humans had withdrawn into their isolated shelters, the sensible robots declared a peace. However, the robots fabricated evidence—films and the like—to keep the humans believing that the dreadful war continued.

Dick is perhaps best remembered for his novel, *Do Androids Dream of Electric Sheep?*, which was made into a movie, *Blade Runner*, starring Harrison Ford. Another of his well-received books was the provocative, *We Can Make You*, which told of the manufacture of celebrity replicas. The cover art of the DAW Books paperback depicts an eye-catching Abe Lincoln robot replica with blank eyes and sophisticated circuitry in its torso.

D. F. Jones, who wrote *Colossus, The Fall of Colossus,* and *Colossus and the Crab* certainly deserves mention. His books told of Charles Corbin, a scientist in charge of constructing the world's most powerful "colossus" computer. The computer turns out to have a mind of its own, however, and decides to become man's ruler. Jones's works were made into an excellent movie in 1970 called *Colossus—The Forbin Project*.

A 1981 novel by Tanith Lee, *The Silver Metal Lover*, covered ground that has been trod before—namely, romance between robots and humans. However, Lee's writing style brings a new dimension to this familiar topic. Her story, set in the future, is about a male robot named Silver. He is an auburn-haired, silver-skinned fellow who plays the guitar. His sensitivity and intelligence capture the heart of a young woman who, unfortunately, violates the moral and social code of her day by consorting with a machine-person.

The following excerpt from Lee's book conveys its theme:

"Mother, I am in love."
"Are you, darling?"
"Oh, yes, Mother, I am."
"With whom, dear?"
"His name is Silver."
"How metallic!"
"Yes. It stands for Silver Ionized Locomotive Versimulated Electronic Robot."
Silence. Silence. Silence.
"Mother…"

John Sladek, a modern-day British author of *The Steam Driven Boy* and other works, takes the story of historical robots and humorously brings them up-to-date. His 1985 title, *Tik-Tok* (after the *Oz* robot character), is a satiric story of the antic adventures of a robot who has a malfunction in his "Asimov circuits" and is no longer obeying the Three Laws of Robotics, promulgated to prevent robots doing harm to humans. Tik-Tok leaves a trail of corpses as he makes his way to high political office in the United States.

Danish author Niels E. Nielsen's 1970 novel, *The Rulers*, took up a more sympathetic theme: the civil and "human" rights of robots. Nielsen's androids demand their rights from their masters and, when refused, go to war in that cause. The wartime conduct of the human race is abominable; they, after all, consider the robots to be mere machines.

Here are more recent entries in robotic or artificial intelligence science fiction:

Neuromancer, by William Gibson
Hyperion Cantos, by Dan Simmons
The Alchemy of Stone, by Ekaterina Sedia
Rainbows End, by Vernor Vinge
War With the Robots, by Harry Harrison
The Ware Tetralogy, by Rudy Rucker
He, She, It, by Marge Piercy
Virtual Girl, by Amy Thomson
The Night Sessions, by Ken MacLeod
Saturn's Children, by Charles Stross

As writers such as Jones, Sladek, Lee, and others prove, the golden age of robotics in science fiction has a more than rosy afterglow. Indeed, we may be on the very dawn of an upswing in robot fiction as new personal and worker robots endowed with artificial intelligence and unheralded physical capabilities inspire a new generation of writers to imagine and contemplate alternative worlds.

Cinematic Machine Dreams

Producers and directors of movies and television shows have carried on their own sort of love affair with robots. One of the most powerful films ever made during the silent film era was *Metropolis* (1926) and there's not been a decade since that hasn't engendered many such movies.

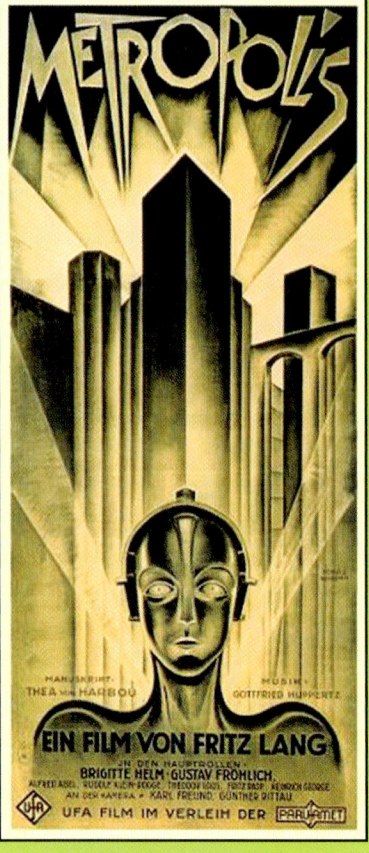

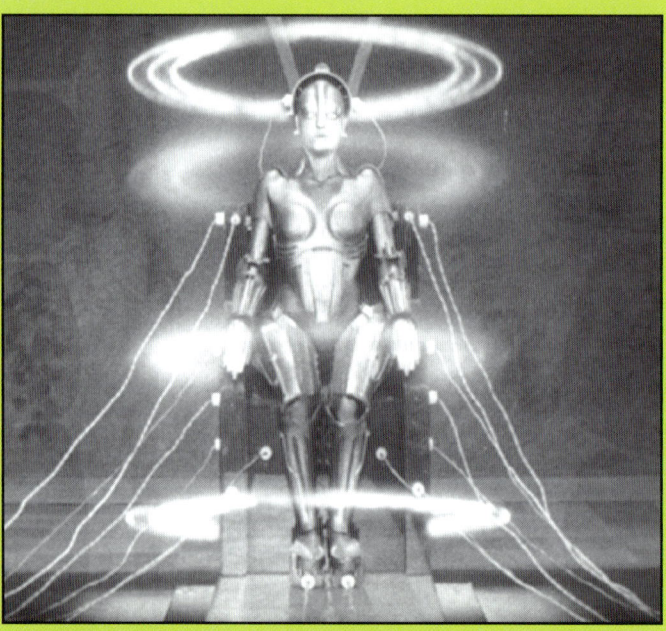

The fascinating and threatening female robot in the classic, 1927 film, *Metropolis*, by director Fritz Lang.

Original 1927 theatrical release poster.

The Forbidden Planet (1956) showcased the beautiful actress Anne Francis. The robot, Robbie, is a spiffy robot character,

The classic *Metropolis*, directed by a German, Fritz Lang, presents an image of a futuristic world in which humanity is firmly split into two classes: the industrialists/owner elite and the downtrodden mass of workers. The beautiful Maria is the leader-heroine of the workers, and an evil scientist, Rotwang, builds a robot that physically duplicates her. The industrialists' plan is to use the robot to inspire the workers to revolt prematurely, giving the state an excuse for harsh repression. Ultimately, their plan fails.

Friendly and Heroic Robots

Just as in literary science fiction, the cinema has alternately depicted the robot as both threat and as faithful servant.

Movies that portray robots in a favorable light include the two classics, *Forbidden*

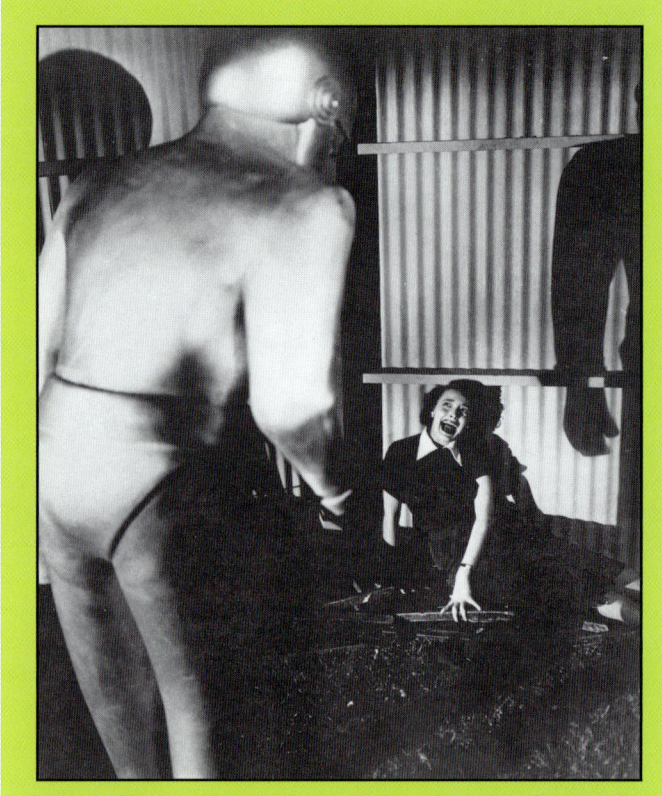

Actress Patricia Neal is threatened by the robot Gort in the movie, *The Day the Earth Stood Still*.

Planet (1956) and *The Day the Earth Stood Still* (1957). In *Forbidden Planet*, Robbie the robot is a wonderful companion to his human masters. He is the prototype of the mechanical household keeper we all long for: a machine that can clean, sew, and cook. Robbie can even manufacture products. Just show him a sample and he'll make you an exact duplicate. Naturally, Robbie speaks; he's cheerful and sometimes very amusing.

In the award-winning *The Day the Earth Stood Still*, the three central characters are actor Michael Rennie, actress Patricia Neal, and the robot, Gort. Gort comes from outer space, but his mission to Earth is laudable. He rounds up all the world's leaders and solemnly announces that, unless they stop warring on each other and establish a permanent system of international peace, he, Gort, will completely destroy the planet. Gort, then, is a true hero—an interplanetary peacekeeper programmed to blackmail the universe—and specifically the factious Earthlings—into doing what they should do on their own.

The Star Wars Saga

The fabulously popular robots R2D2 and C-3PO of *Star Wars* (1977) and its sequels, *The Empire Strikes Back* and *Return of the Jedi*, are the perfect archetypes of the loyal, helper robot. These two androids, one shaped like a trashcan, the other resembling a small, metallic gold man, assist human hero Luke Skywalker as he battles a corrupt galactic empire bent on ruling the universe.

Director George Lucas's *Star Wars* trilogy gives us images of many other animated machines, as well. In one scene, a motley collection of used robots is sold much as we now handle transactions for used cars.

Another movie, *Android* (1982), won critical acclaim, but its commercial appeal didn't approach that of *Star Wars*. Directed by Aaron Lipstadt for Show Films, Ltd., (an Island Alive release), *Android* may well become a robophile screen classic.

In this imaginative picture, Don Opper plays the lead as Max 404, an android created aboard a space station by a cruel and insensitive scientist, Dr. Daniel (Klaus Kinski). Max 404 is an innocent who becomes a great fan of Earth artifacts and culture, including blue-jeans and film legends like Jimmy Stewart and Humphrey Bogart, whom he tries to imitate in dress and manner.

The two loveable robots, C-3PO (left) and R2D2, in *Star Wars*.

Soon, Max 404 is joined by a female android, Cassandra, played by Kendra Kirchner. Meanwhile, a murderer threatens the androids and Max 404 and Cassandra also become aware of the evil designs of Dr. Daniel. All's well in the end, however, as both the scientist and the murderer are destroyed, leaving the two androids free to assume human identities and live happily ever after.

The Robot Will Get You If You Don't Watch Out

Just as they frequently are portrayed in print, some cinematic robots take up roles as compliant tools of galactic bad men, as rebellious servants, and as malfunctioning creatures who run amok.

Most of the films that characterize robots as threats are movies. These include *Gog, Gog the Killer*, and a thirteen-segment Republic Films serial called, *The Vanishing Shadow*. Only a few pictures featuring bad guy robots have won acclaim or recognition as quality productions of enduring value.

Probably best known among the films featuring laboratory-created life forms are many versions of Frankenstein flicks. Over the years, viewers have delighted in or suffered through (depending on one's perspective) such films as *The Bride of Frankenstein, Curse of Frankenstein, I Was a Teenage Frankenstein*, and *Frankenstein Meets the Werewolf*. Finally, in 1969, these films gave way to satire as an East German comedy, *Hollow My Weenie, Frankenstein* made its appearance. It was followed by Mel Brooks's very funny spoof of the entire Frankenstein genre, *Young Frankenstein*, which starred Gene Wilder and Madelyn Kahn.

And More Robot Films for You

We can't list them all, of course, but here's a glimpse through the keyhole at some of the members-in-good-standing whose celluloid images populate the Robot Hall of Fame (or Infamy, take your pick!).

Phantom Empire (1936). Singing cowboy hero Gene Autry is on the side of law and order in this depression-era, silver screen production. The outlaws have robots on their side. Guess who emerges the victor in the ensuing drama?

Satan's Satellites (1951). The evil robots in this film serve aliens from outer space. But Commander Cody comes fearlessly to the rescue.

Lovely ladies were often featured in the early cinematic productions.

Robot Monster (1958). This movie is said to be one of the silliest and most improbable ever produced. Its plot, which includes robots who wear gorilla suits and who come out of the ocean to attack women sunbathers, may just earn a cult following for this campy film.

2001: A Space Odyssey (1968). HAL is the computer brain of the robot spaceship in this Stanley Kubrick recreation of the best-selling book by Englishman Arthur C. Clarke. HAL is programmed to be logical—too logical. To insure the success of the mission, he methodically destroys the human beings aboard the spaceship. However, in the sequel, *2010*, HAL redeems himself by self-sacrifice.

The Stepford Wives (1974). Perfect simulations of housewives are built. They murder their real-life models and become obedient, traditional wives to macho men.

Demon Seed (1977). In this MGM production, beautiful Julie Christie plays a woman who is enslaved and impregnated by a terrifying, machine-like robot who knows the meaning of lust.

The Black Hole (1979). The proverbial mad scientist, in this case a Dr. Durant, uses hulking mechanistic robots to accomplish his diabolical plans.

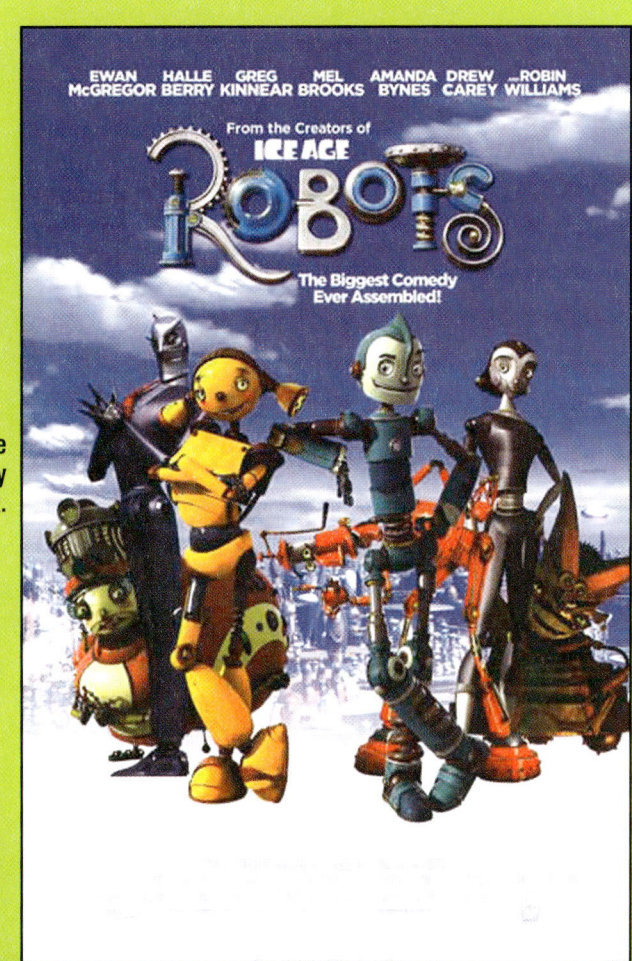

Robots is a 2005 comic science fiction file produced by Blue Sky Studios for Twentieth Century Fox.

Actor Robert Downey Jr. in the movie, *Iron Man* (2013).

Robots on Television

TV productions with robots in the cast include two British series: the long-running "Dr. Who" and "The Avengers." "Dr. Who," also shown in America, features robot dog "K-9," and the evil Dalek race of machine creatures.

"Riptide," a popular NBC detective series by Stephen Cannell Productions, displayed a strange-looking but friendly mobile robot named Roboz. Murray Bozinsky, the show's resident computer hacker, affectionately called the robot that "squat, ugly, orange thing!" The mute Roboz did not have many capabilities, but he added a touch of high tech class to the show.

Three other TV series with robots and near-robots are the revived "The Jetsons," the continuing story of a twenty-first-century family, their dog, Astro, and robot maid, Rosie; "Small Wonder," a syndicated situation comedy about a cute little bionic girl who's adopted by a typical family; and "Automan," a fantasy about an animated hero created out of thin air by a computer.

Most robots appearing on television have been at least benign, the exceptions being evil robots who appeared in a few of the "Star Trek," and "Battlestar Galactica" episodes. The "My Living Doll" series (1964) starred a likeable female robot as did the 1976 series "Holmes and YoYo." In "Lost in Space," a very popular mid-'60s adventure about stranded space travelers, the unnamed robot was a brainy, sensitive, and kind mechanical android who constantly kept the bumbling, insensitive—but laughable—Dr. Smith on the straight-and-narrow.

In the 1980s "Knight Rider," a principal character is a robotic automobile named K.I.T.T. The vehicle is packed with electronic gadgetry and endowed with a pleasant voice. In the series, K.I.T.T. and a human companion battle an assortment of seedy criminals.

Mere Dreams No Longer

As one reads of robots in early science fiction novels and pulp magazines, and as we view gripping action adventures of robots in past movies, a stunning realization takes hold. Modern technology and science have in many cases outstripped past fantasies; many of the dreams of science fiction prose have become reality. It appears that the boldest and most imaginative of robot tales—even some of those that have appeared the most improbable—may yet come to pass.

The robot breed's evolutionary process is gathering momentum as we move swiftly into the twenty-first century. Already we have put robots in space, and—shades of Frankenstein!—organ transplants have become commonplace. We have robots in our homes that talk and walk and robots in our factories that can work faster and with more precision and endurance than human workers. Science fiction has become science fact.

How we have arrived at this juncture is the subject matter of the next chapter. But first, let's take a fun journey down memory lane as we review the Golden Age of robots and androids in science fiction. These pictures are only a few of the countless magazine covers we have encountered, each of which provides an imaginative look at our future.

The Golden Age of Robots and Androids in Science Fiction

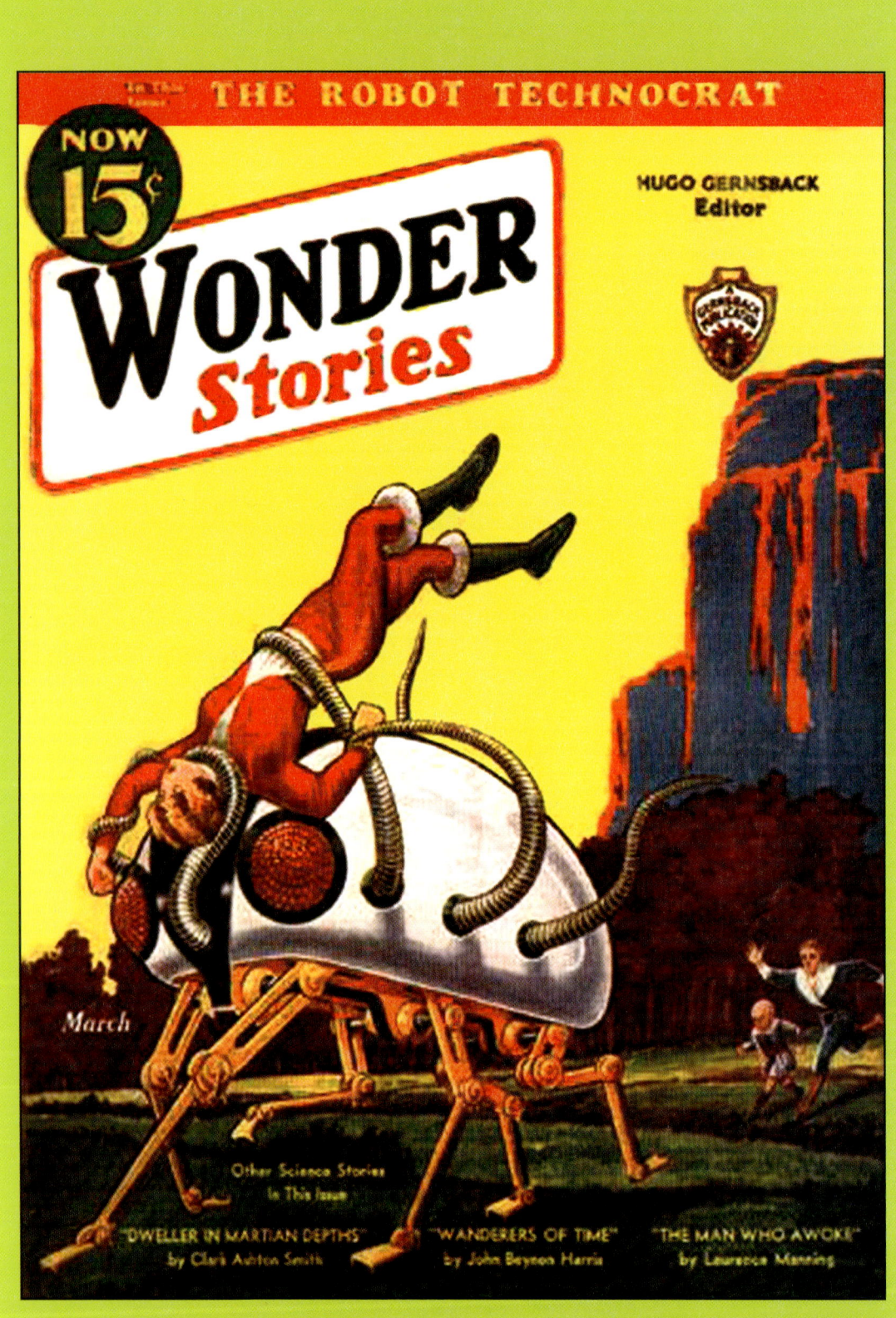

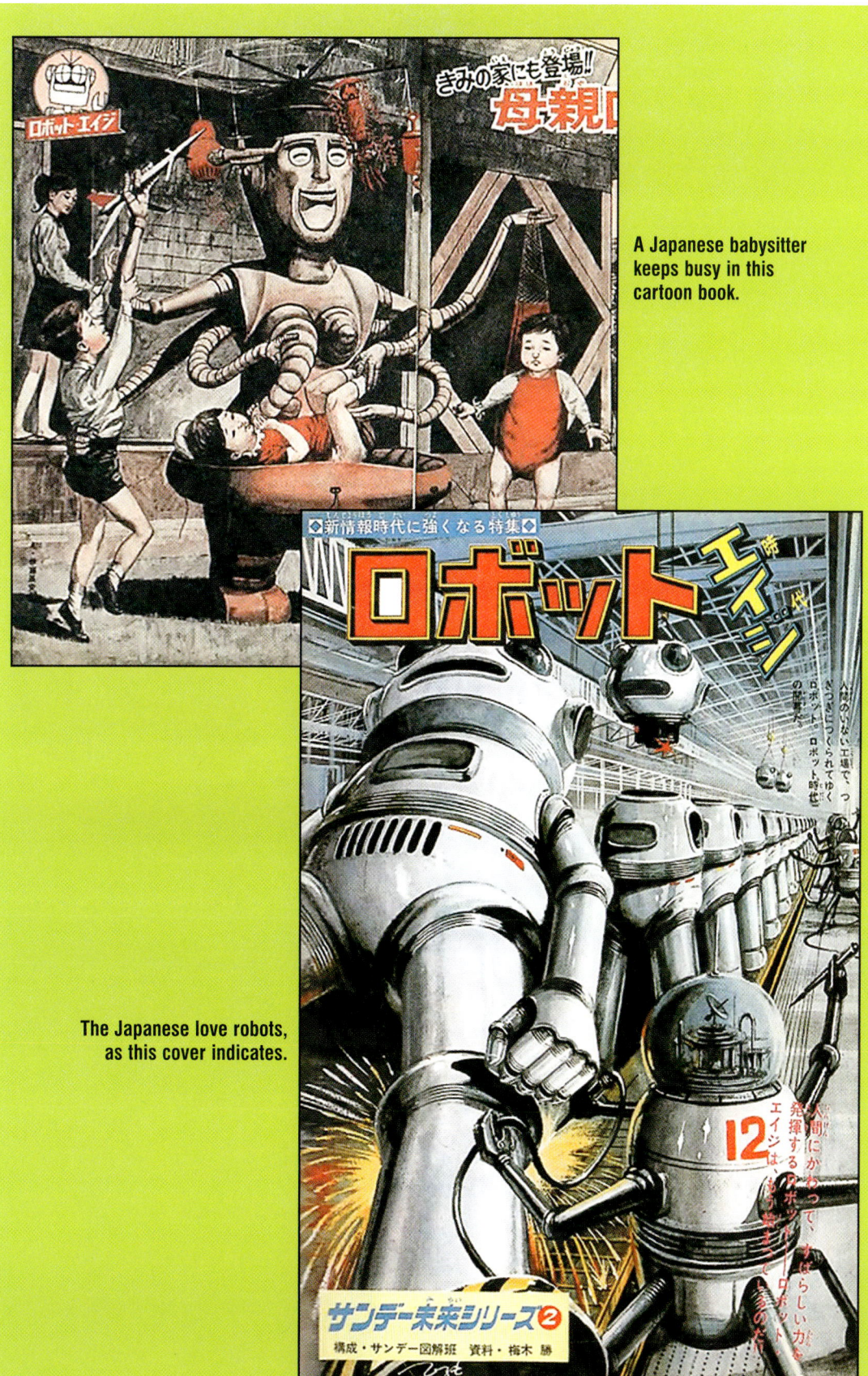

A Japanese babysitter keeps busy in this cartoon book.

The Japanese love robots, as this cover indicates.

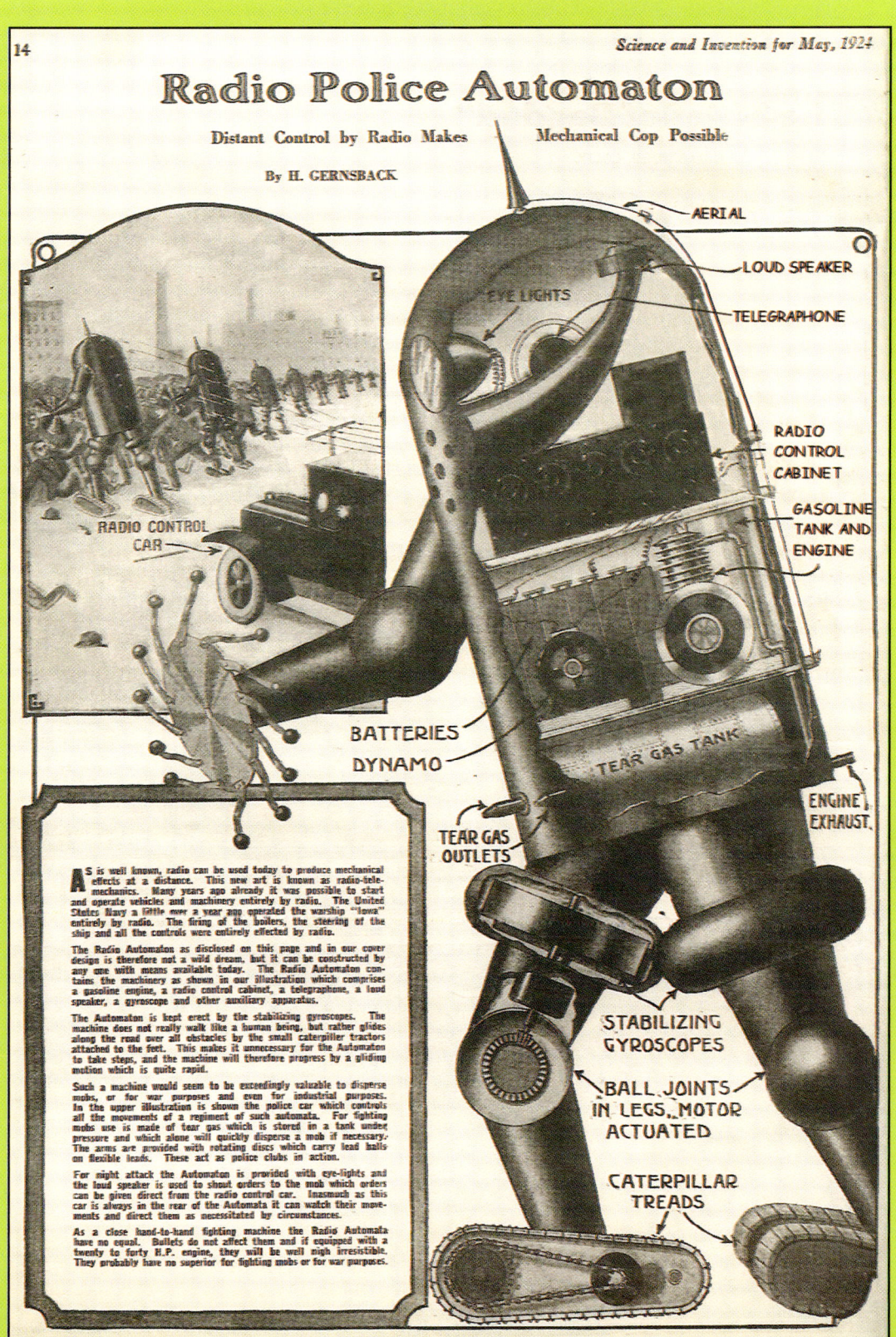

A police robot in *Robocop* (1924).

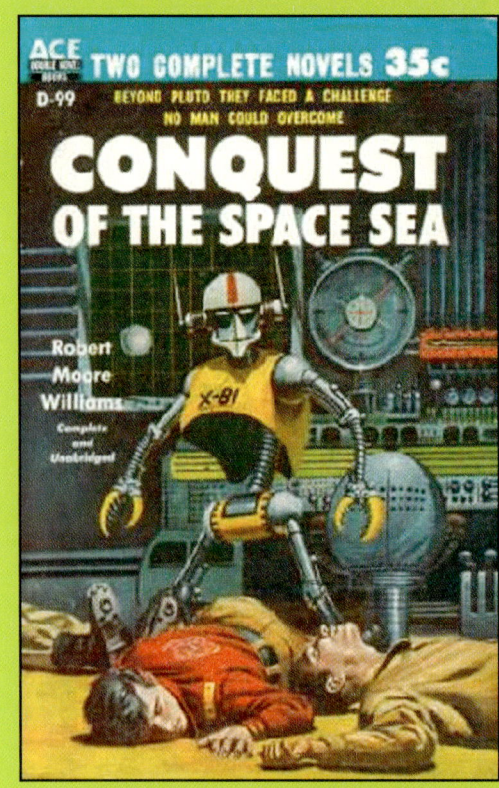
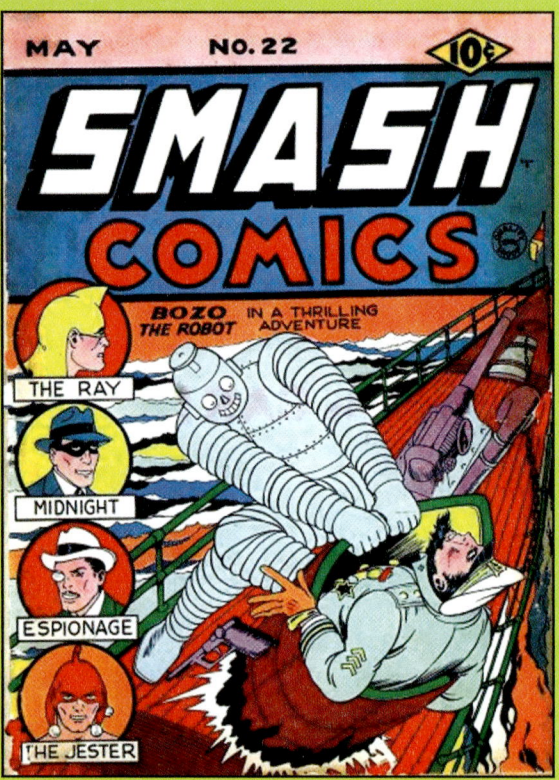

The Mechanical Man (1924) is an early Italian sci-fi film. In the film, a criminal leader kills the robot's scientist inventor, then uses the robot to commit a series of crimes. Fortunately, the inventor's brother builds another robot to challenge the Mechanical Man. The two fight to the death in an opera house and the criminal is electrocuted.

Above: *Conquest of the Space Sea* is about a robot warrior, X-01.

Top, right: Bozo the robot in a thrilling adventure.

Chapter 9

Robots: Be Part of the Beginning

"Ours is truly the first generation to experience the birth of a new life form, however primitive it may now be, and no tale in "Ripley's Believe It or Not" or in a science fiction thriller could be more exciting and breathtaking than the pleasure of taking an excursion down the robot road to the present.."

—Texe Marrs
The Personal Robot Book

"Be part of the beginning!" invited the promotional literature for the 1984 Albuquerque, New Mexico, First International Personal Robot Congress and Exposition. The brochures proclaimed the arrival of the Robotics Age and invited enthusiasts to come celebrate the momentous event. And come they did, to hear Isaac Asimov speak from New York City via a satellite link, and to meet personal robot entrepreneurs such as Joe Bosworth of RB Robot Corporation and Nolan Bushnell of the fledgling Androbot, Inc. The attendees also met several score robots and enjoyed the promotions of exhibitors offering not only complete machines but robot parts and software and books about the creatures.

The industrial robots also have their own show, staged each year in a large U.S. city by the Robotics Industries Association and the Society of Manufacturing Engineers. At this convention hundreds of sales representatives approach potential customers eager to observe the latest technological advancements in worker robotics.

These key robot shows tell us only the latest chapter of what is already a long story. At these trade events we are able to observe the results of nearly 5,000 years of technological progress in robotics. We can roughly divide the natural history of robot evolution into five time frames. During each of these eras, significant advances were made that set the stage for subsequent progress.

The Dawn of Robotics (3000 B.C. to A.D. 1199)

From about 3000 B.C. to A.D. 150, craftsmen in many parts of the world created wood

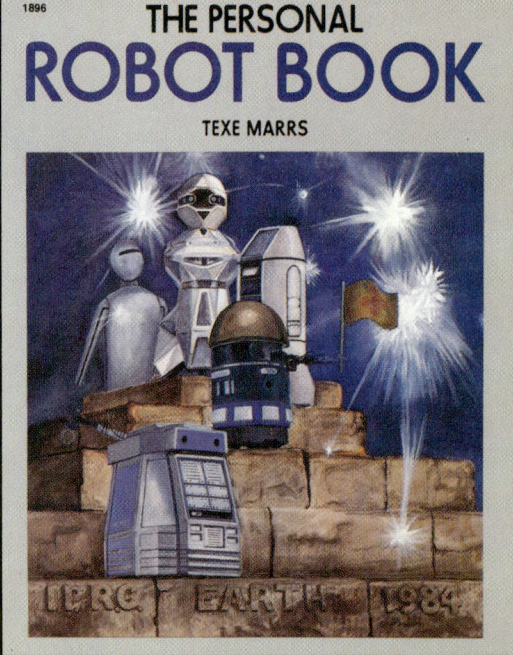

The Personal Robot Book, by Texe Marrs, was the first ever book to survey the field of personal and home robotics. Many of the robots it describes were on display in Albuquerque, 1984.

Two young boys are puzzled by a moving, talking robot at the International Personal Robot Congress and Exposition in Albuquerque.

and metal figures, toys, puppets, and talking heads that were operated by primitive gears, string pulls, levers, fulcrums, axles, water power, and gravity. These objects were built primarily for the amusement of royalty, to celebrate special events, or for use in religious ceremonies and other rituals.

In Africa, inventive priests built lifelike idols to be worshipped by superstitious tribes-people. One such creation employed a draw cord attached to the idol's jaw to cause its mouth to open; a speaking tube or ventriloquist endowed the idol with speech.

In imperial Rome, priests and priestesses used string-controlled puppets to prophesy and to abet their own aura as the earthly representatives of the gods.

In the third century B.C., Greek thinker Archimedes set forth theories about the use of steam for power. From about 200 to 100 B.C.—in Egypt, Ethiopia, Persia, and China, engineers created automatons of animals, birds, and people that operated by steam. Building on Archimedes' work, Hero of Alexandria (circa A.D. 100) wrote about the mechanics of automata. Hero proposed that pneumatics (air pressure) could be used to power machines and he outlined the principles of the crank, the screw, the cogwheel, the camshaft, the pump, and the piston. Hero also invented a heat transfer system that caused doors to open automatically.

In the fourth century A.D., Chinese builders erected a golden statue of Buddha set on a carriage on which were mounted animated figures resembling Taoist monks. As the carriage was drawn, the monks revolved around the statue, bowing and depositing incense into a censer. Later, in 790, a Chinese inventor constructed a wooden otter that could actually catch fish. The next century, in 890, another Chinese craftsman built a wooden cat that, observers reported, could catch rats. Meanwhile, in nearby Japan, The

Prince Kaya supervised the construction of a mechanical doll that could raise a bowl of water and pour the water over its own face.

The Age of Automata (1200 to 1821)

After a European hiatus of many hundreds of years, creative minds there labored to build automatons that could move as if they had life. From A.D. 1200 to about 1821, many mechanical automata were constructed, some after the human form, others in the shape of animals. Among the notables involved in the quest to create mechanical life were Bavarian Albertus Magnus, Englishman Roger Bacon, and the Italian Leonardo da Vinci.

Magnus (1193-1280) constructed the servant automaton mentioned earlier, which had a lifelike appearance, supposedly possessed faltering speech, and was able to open doors for guests. Bacon (1214-1294) was reputed to have created a talking head that was more than seven years in the making. The renowned inventor-artist, da Vinci (1452-1519), was inspired by the visit of King Louis to build a mechanical lion to honor his majesty. Entering Milan, the king was astonished to see the lion stealthily approach him. However, his astonishment turned into amusement and mirth when the lion suddenly stopped, opened its chest with a yank of its paw, and pointed to the coat of arms of France emblazoned there.

Beginning in the fourteenth century, workmen and inventors also began to build mechanical dolls, statues, figurines, and clocks. Automatons in the shape of animals and birds were especially popular, and nobility and the clergy requested that such devices be installed by architects and builders of civic buildings and churches. An example of such handiwork is a fourteenth-century mechanical crowing rooster perched atop the cathedral of Strasbourg, France. Each noon, the metal rooster flaps its wings and sticks out its tongue. With minor repairs over the centuries, Strasbourg's mechanical rooster continues in its task.

In the eighteenth century, European automaton-makers Baron Wolfgang von Kempelen and Jacques de Vaucanson gained a measure of fame. The latter built a realistic mechanical duck that "chattered,…swam, splashed in water, and…spread its wings." The fowl could smooth its feathers and swallow kernels of corn fed to it by hand. Vaucanson also built a tiny walking mechanical asp for us in an eighteenth-century production of *Cleopatra*. Baron von Kempelen was the creator of talking machines as well as of the famed "Turkish Chess Player."

The Jacquet-Droz Craftsmen

Perhaps the most famous of eighteenth-century makers of automatons were the father and son team of Pierre Jacquet-Droz (1721-90) and Henri-Louis Jacquet-Droz (1752-91). Assisted by a very talented mechanic, Jean-Frederic Lescho, the Swiss-born Jacquet-Droz proved they were master craftsmen of unparalleled skill and vision by constructing a number of remarkable, humanlike automatons.

In 1774, the Jacquet-Droz showed to the public three of their finest creations. The Scribe (also called *The Writer*) is a "child" about three years old. In his right hand is a goose quill. When he writes, his eyes follow the tracing of each letter. *The Draughtsman* can execute a portrait of Louis XV, while *The Lady Musician* is a charming young lady who graciously plays melodies on a small pipe organ. The Jacquet-Droz figures are now

The Lady Musician is a charming young lady who plays melodies on a small pipe organ. By the Jacquet-Droz father and son duo.

The Scribe, an 19th century automaton by mechanics and artists Pierre and Henri-Louis Jacquet-Droz

on permanent display at the Musée d'Art et d'Histoire in Neuchatel, Switzerland.

The fame of the Jacquet-Droz spread throughout Europe, encouraging many other clockmakers and mechanics to try their hand at building complex automatons, most of which were sold to the wealthy and the nobility. One such person was Henri Maillardet, an apprentice under the Jacquet-Droz. Maillardet's impressive *The Writing Child*, built in 1815, is now on display at the Franklin Institute in Philadelphia, France's King George III presented a Maillardet automaton doll to the Emperor of China as a gift. This exquisitely crafted doll wrote its words in Chinese calligraphy.

The Age of Electricity and Machines (1822-1920)

Insofar as it relates to robotics, the Age of Electricity and Machines begins in 1822, when Englishman Michael Faraday (1791-1867) invented the electric motor. However, even before Faraday's innovation there were a number of other discoveries of great significance to the development of practical robots.

James Watt's steam engine (circa 1782) gave some promise of powering automatons. American Oliver Evans also was a pioneer in the development of steam engines. In 1783 he built a flourmill in Philadelphia that was almost totally automated.

Other innovations of significance to robotics during this period included the mass production in France of the Jacquard loom, a machine programmed by punched cards (1801), and the development by American Christopher Spencer of the automat, a cam-programmable lathe (1830). Further foreshadowing the future of robotics and automation, in 1892 Spencer Babbitt designed a motorized crane and gripper to remove hot ingots from steel furnaces.

Several nineteenth-century inventors attempted to put steam to use in powering automatons. One such device was built by George Moore in 1893. This fascinating steam robot was propelled by a half-horsepower motor that caused jointed rods to move the robot's legs. Exhaust pipes protruded from the robot's mouth and head.

From 1822 to 1906, technological advancements using electricity came with increasing regularity. In 1838 the world saw the Samuel Morse telegraph; in 1876, Alexander Graham Bell gave us the telephone, and in 1877 inventor *par excellence* Thomas A. Edison presented his phonograph machine. Then, only two years later, Edison's brilliance was again demonstrated by his invention of the first commercially operable electric light bulb.

Edison also gave the world another invention: a talking doll (circa 1894). The doll,

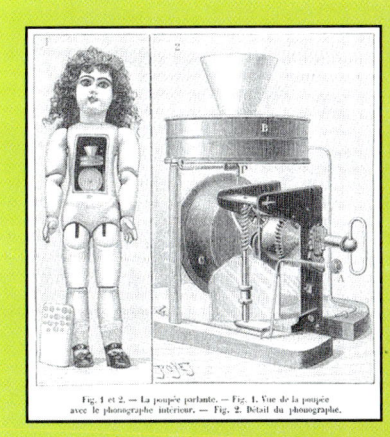

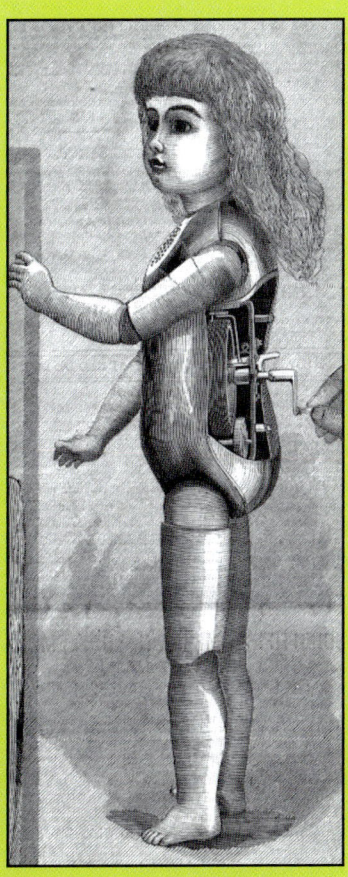
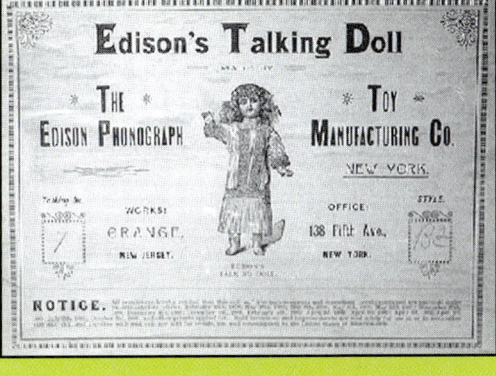

Thomas Edison's "Talking Doll" (circa 1890) was very popular.

which by the use of a crank and a phonograph cylinder was able to recite the nursery rhyme, "Mary Had a Little Lamb," caught the public fancy, mainly because of the popularity of its inventor. Soon, Edison's factory in New Jersey was producing and selling five hundred of the dolls per day.

A few years earlier, in 1877, U.S. inventor J. D. Hughson patented an automaton more useful than Edison's faddish doll. His device was an electric railroad signal in the shape of a man wearing a top hat. Hughson's automaton wasn't a commercial success, but its invention proved a significant step toward applying the technology of automata for purposes other than entertainment.

American immigrant Nikola Tesla (1856-1947) was also a pioneer in the field of electricity and robotics. The inventor of the alternating current electric motor, Tesla worked feverishly on electric-driven robotic devices. In 1898, to the utter amazement of thousands of people gathered at Madison Square Garden, Tesla demonstrated a remote controlled electric submersible boat.

Tesla's ultimate goal, never achieved, was to build machines that possessed intelligence. Tesla wrote: "I think the time is not distant when I shall show an automaton which, left to itself, will act as though possessed of reason and without any willful control from the outside." Also, we come upon his bold prediction: "Teleautomata will ultimately be produced, capable of acting as if possessed of their own intelligence, and their advent will create a revolution."

The Android Period (1921-1944)

Borrowing from the works of Faraday, Edison, Tesla, and others, and inspired by the coining of the term "robot" in 1921, inventors soon began to install electric motors and electrical relay devices in their mechanical creatures. These electrical systems were joined with sophisticated mechanisms of pulleys, belts, cables, gears, shafts, rollers, and levers to produce incredible new androids. They created a sensation wherever they were displayed, which was exactly the purpose of their creators, companies such as Westinghouse, which sought to show the public the ultimate possibilities of electricity.

In Great Britain in 1928, an "aluminum man" that could rise, bow, and make a speech was put on display at the Royal Horticultural Hall in London. Recognized as the very first British robot, spectators dubbed it the "Royal Knight."

From 1927 to 1940, the Westinghouse Corporation presented three generations of mechanical men. In 1927, Televox was born, followed by Willie in 1931 and Elektro in 1939. Elektro had a pal, a robot dog named Sparko, built in 1940.

The interesting story of Elektro and Sparko began when Westinghouse engineer J.M. Barnett built Elektro as a publicity gimmick for the 1939 New York World's Fair. Elektro had a bag of twenty-six tricks. He could walk forward and back, bow his head, and crane his head at a crowd. When in the mood, he brought his hand to his head in a patriotic salute. The robot could count, one finger at a time, and distinguish between the colors of red and green. He also was a fancier of cigarettes and cigars.

Elektro weighed 260 pounds and had 48 electric relays inside his huge body. The robot's inventor, Barnett, noted that Elektro was not capable of completely duplicating all the human body's movements because to do so "his brain would have to contain 1,026 electric relays, weigh approximately 1,000 pounds and occupy about 108 cubic feet of space."

Nikola Tesla was one of the world's greatest inventors.

In 1898, Tesla demonstrates his "teleautomaton" remote-controlled boat at Madison Square Garden, New York.

Elektro had no remote control, instead responding to voice commands using a telephone handset connected to its chest. The chest cavity even lit up as it recognized each word. Each word set up vibrations which were converted into electrical impulses, which in turn operated the relays controlling eleven motors.

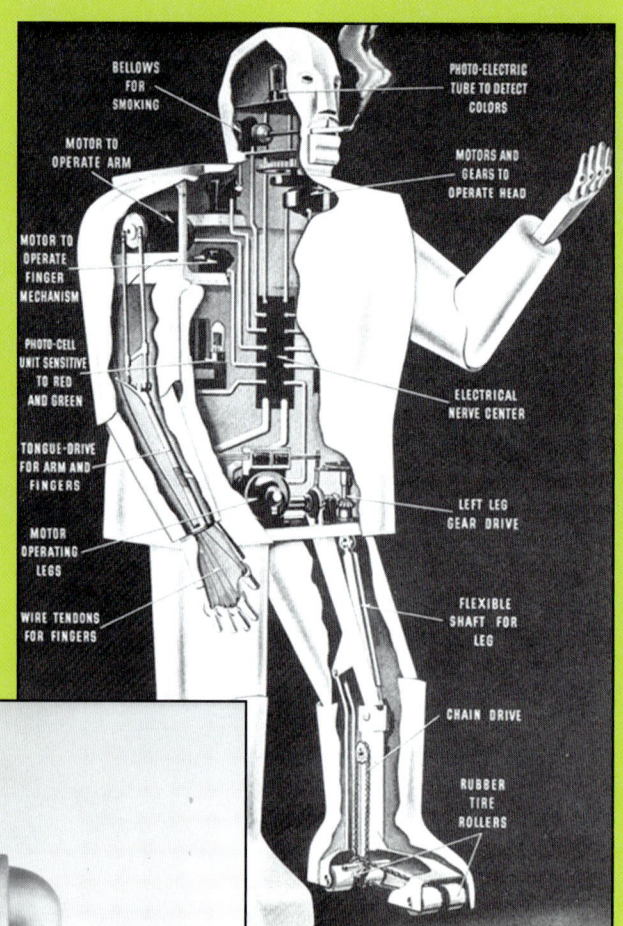

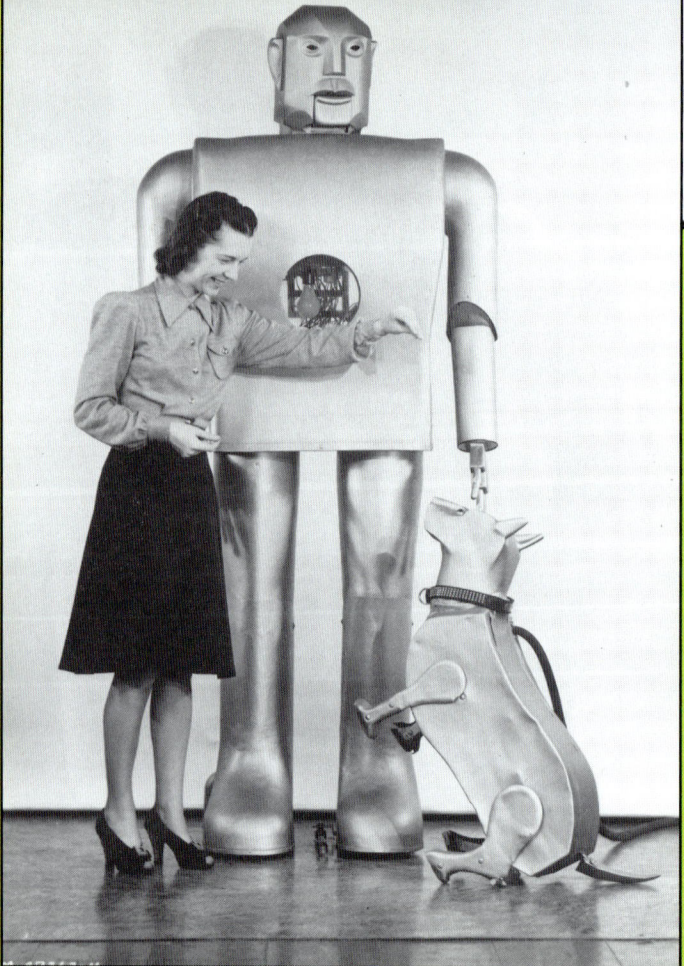

Elektro, the Westinghouse automaton, with a model (left) and his dog, Sparko, were on exhibit in 1940 at the New York World's Fair.

According to Barnett, there wasn't "the remotest possibility that such a gargantuan metal man will ever be built." Added the inventor, "No engineer would ever be so ridiculous as to imagine that any robot could ever take the place of man."

Given the impressive accomplishments of Elektro in 1939, one wonders what a top-notch robot builder like Barnett could do today, using the marvels of the electronic computer, the microchip and the biochip as building blocks.

The Computer and Robot Revolution (1944 to Today)

Today, less than fifty years after Elektro, science has provided robot brains, each so small that *several thousand* of them could fit into the 108 cubic feet of space Barnett said would be required. The microchip, a tiny sliver of silicon and integrated circuitry, is the heart of the modern electronic computer. For some mental processes—such as calculating and recall—contemporary computers are superior to the human brain. But in other respects, particularly in the realm of reasoning and logic, computers are woefully deficient compared to the processes of the biological human brain.

Still, in another 25 years, as molecular computer chips, neural implants and high density microchips are manufactured, the computer brain of the robot may well be the equal of or even surpass that of human beings. Moreover, that superior machine brain may be much, much smaller than its human counterpart.

The Computer Revolution and the Microchip

The Computer Revolution and the Robot Revolution have developed in tandem. The computer epoch began in earnest in 1944 when an American graduate student, Howard Aiken, invented the first digital computer, Mark I, a behemoth of a machine that used electromagnetic relays. Two years later, an U.S. Army research team at the University of Pennsylvania demonstrated ENIAC (Electronic Numerical Integrator and Calculator). This first, all-electronic digital computer was the precursor to today's sleek, powerful—and much smaller—models.

ENIAC was so big that it filled a large, barn-sized room. It had 17,000 bulky vacuum tubes. But in 1948, William Shockley, a Palo Alto researcher, patented his invention, the transistor, a small electronic component that soon made vacuum tubes obsolete and initiated the

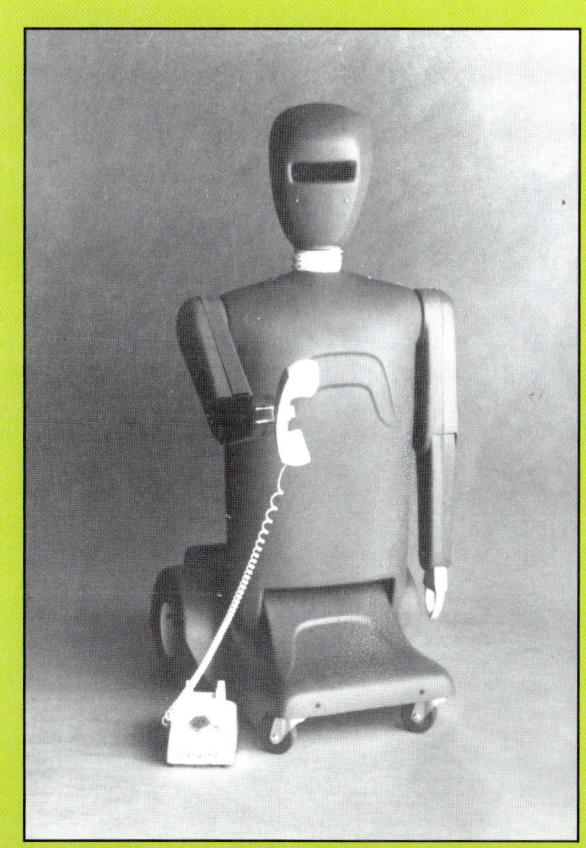

Marvin was "born" in 1985. His father was Iowa Precision Robotics, Ltd. Marvin moves on his wheels, speaks 500 words, moves his head and is controlled by computer.

Cyber I has advanced artificial intelligence and machine vision which helps him avoid obstacles in his path. (Cybernetics)

process of computer miniaturization that continues today.

Also in 1948, scientist Norbert Weiner, founder of cybernetics, published his book, *Cybernetics: Control of Communication in Animal and Machine*. Weiner's theory suggested that the computer and the human brain were somewhat comparable in that each employed systems of feedback and memory. His fresh way of thinking about machine intelligence marked a turning point in robotics.

Five years later, in 1953, English scientist W. G. Walter's important and provocative book, *The Living Brain* (Duckworth, London) proposed that machine species could be classified much as are biological species.

Another breakthrough in computers and robotics occurred in 1958 when Texas Instrument's Jack Kilby invented the integrated circuit on a semiconductor wafer of silicon: the microchip. A year later, Robert Noyce improved upon Kilby's microchip and the wheels were set in motion for the Computer Revolution. Transistors gave way to microchips, and mainframe computers to micro-computers and even pocket-size computers. All that was necessary was to marry the computer to the robot.

H.E.N.R.Y., built by Bruce C. Taylor, uses bumpers to stay clear of obstacles. This robot possesses a 128-word vocabulary and has two workable arms. (theoldrobots.com)

Robots on the Assembly Line

While researchers were busy miniaturizing computer components, robot builders weren't standing still. In 1954 George C. Devol, a Kentucky engineer and entrepreneur, patented a robot, called "Unimate," for use on assembly lines. It was the first commercial robot offered by Devol's fledgling company, Universal Automation (later, Unimation). Devol went on to patent thirty-nine other robotic products, but his robots seemed like such a departure to factory managers of his time that few of them were sold.

Disappointed, Devol sold his patents to Consolidated Diesel Corporation (Condec). Enter Joseph Engelberger, an enthusiastic and visionary engineer who took over Unimation and ran with it. Fighting resistance from industry, the optimistic Engelberger barreled ahead. In 1961 the first Unimate robot was installed on a General Motors assembly line. This first-generation machine was only a "dumb," automatic, pick and place robot, but its employment signaled the beginning of the Robotics Age. Industry was changed forever.

Robots Come Home

The Computer Revolution also energized the development of the home and personal robot. The first mobile robot controlled by a computer brain was Shakey, developed in 1968 by SRI International of Menlo Park, California. SRI's Charles Rosen endowed Shakey with the ability to wheel around the laboratory, see with television eyes, and make modest decisions about obstacles in his path and how to avoid them. Shakey was the forerunner of what will undoubtedly become hundreds of thousands—perhaps millions—of home robots.

In the 1980s, sensing that the time was ripe as evidenced by the ongoing craze for personal computers, a half-dozen companies in the United States introduced mass-produced computerized home robots to the marketplace. Foremost among these firms were Androbot, headed by Atari Computer founder Nolan Bushnell, and RB Robot, a small Colorado corporation led by Joseph Bosworth. The Heathkit Company, a Zenith subsidiary, also produced a personal robot, HERO 1, that doubles as a robotics educational system.

In 1985, I wrote my groundbreaking book, *The Personal Robot Book* (McGraw Hill/Tab), which became an instant classic and was chosen by the computer and electronic book clubs. The following year, I authored the first book ever on robotic jobs, *Careers With Robots* (Facts on File).

The arrival of the home robot and the introduction of worker robots to replace human workers is fulfilling the age-old dreams of visionaries, scientists, and philosophers. Surely, there are few technological parallels with this achievement of the invention of machines that replicate human activity. Aristotle, Archimedes, Hero, and even Edison would no doubt stand in awe of today's robots, were they to be brought to life for an instant to glimpse what their ideas and labors have helped bring to pass.

Chapter 10

Humans Become Cyborgs—The Bionic Human

The use of artificial limbs and body replacement parts has been common for many centuries. In the 1570s, Ambroise Pare, a French surgeon, developed a variety of prosthetic devices for crippled persons. And in the United States, in the nineteenth century, many an artificial leg and arm was sold by mail order. Soon after his transatlantic flight in 1927, aviator Charles Lindbergh and a colleague invented an external heart pump and demonstrated it to the public. However, all of these prior achievements pale when contrasted with the incredible array of artificial limbs, replacement parts, and bodily function surrogate equipment now in use or under development.

Today, we call this field *bionics*, (also *biomimetics*) indicating the combination of biology and electronics. Bionics does concern itself with the biology of the human body and with electronics, but more broadly, the field encompasses all the facets of medical technology concerned with bodily functioning. This includes not only spare-parts medicine to replace damaged body parts with artificial and natural substitutes but also the use of computerized sensors, prosthetic devices, and electromechanical machines.

Biotechnologists, biomedical engineers, and bionics technicians are working with specialists in neurology, chemistry, pharmacology, medicine, electronics, ceramics, mechanical engineering, and other fields to build artificial human parts. Bionics

Discovered near Luxor, Egypt, these fake toes are the earliest prosthetics found. They are now in the British Museum.

professionals also develop new composite synthetic materials, construct computers and microchip sensors that simulate and calibrate bodily functions, and develop fiberoptics, lasers, and biotechnology as well as other advanced technologies for application to bionics.

Bionics Advances

Bionics technology is progressing rapidly. The Six Million Dollar Man—the bionic human—is edging closer and closer to reality as scientists, engineers, medical doctors, and technicians team up. We now have artificial substitutes for the heart, ear, lung, kidney, pancreas, joints, and external limbs. And almost every other body part is being researched in the hope that workable artificial replacements can be developed. We have also produced artificial neurons and neural networks. When we examine just a few of the fascinating new developments in bionics we realize how dramatic and revolutionary this field is. A look at the state-of-the-art in bionics is also relevant to the future of robots and robotics.

THE ARTIFICIAL EAR

The artificial ear is here. A cochlear implant, it may eventually restore or bestow the gift of hearing on as many as one-third of all deaf people, many of whom are profoundly deaf. Symbion, the same medical research firm that offers the Jarvik-7™ artificial heart and other organs, manufactures such an implant, called the Ineraid™.

The Ineraid™ consists of two parts: an implanted electrode assembly and an external microphone with sound processor. Surgery is required to thread the electrodes into the existing cochlea. The sound processor box is carried in the person's pocket or worn on the belt.

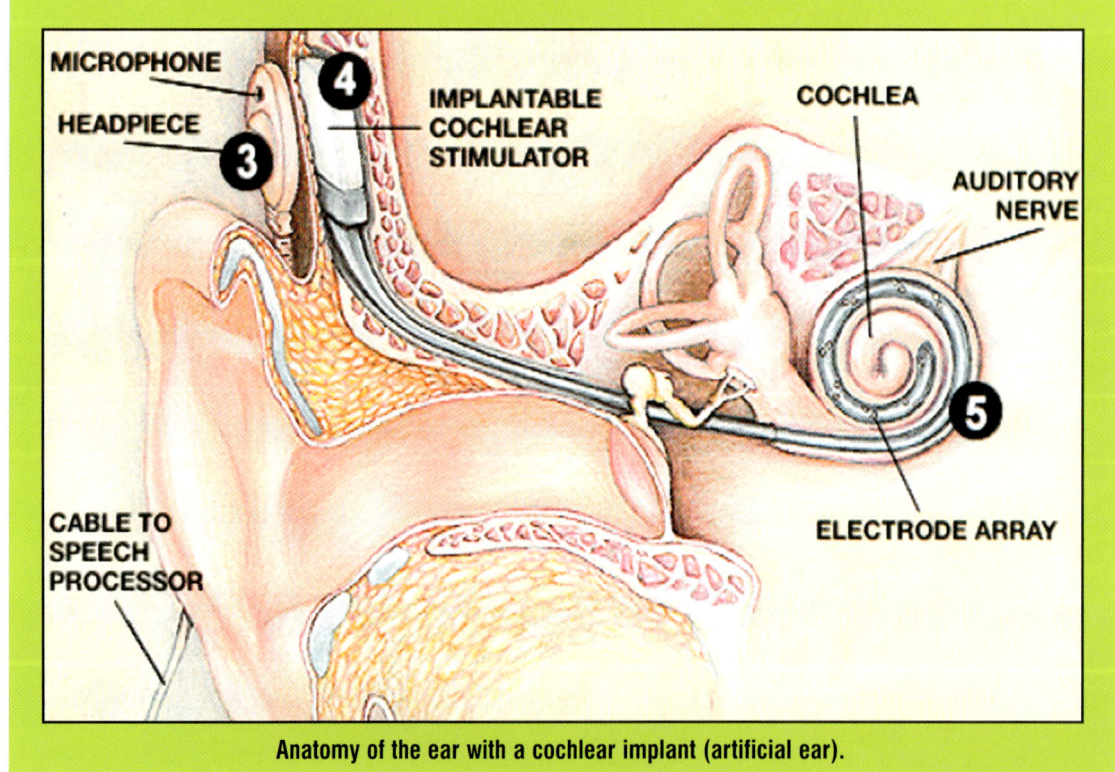

Anatomy of the ear with a cochlear implant (artificial ear).

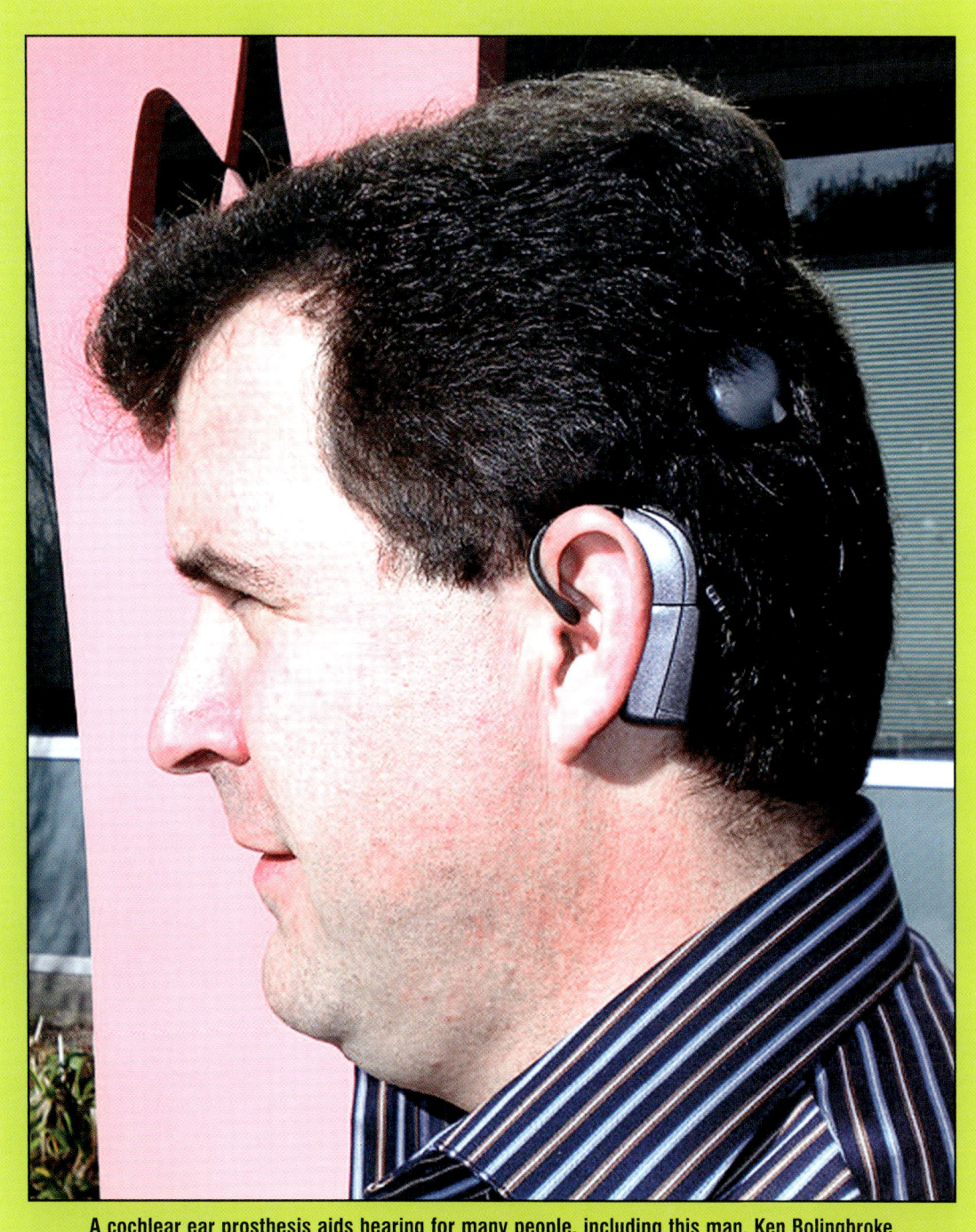
A cochlear ear prosthesis aids hearing for many people, including this man, Ken Bolingbroke.

The artificial ear works when sounds entering the microphone are relayed to the sound processor. These sounds are then transmitted to the electrodes in the cochlea, which imitate the function of the damaged biological cells. This electrical information is transmitted by the auditory nerve to the brain where the sounds are interpreted as meaningful information.

ARTIFICIAL EYES

The first bionic eyes are here, and some blind people have limited sight.

Ms. Diane Ashworth, of Australia, was totally blind when she had this bionic eye implanted. "Suddenly, I could see a little flash of light," she said. "It was amazing."

The artificial heart has saved and extended many lives.

ARTIFICIAL HEARTS

The artificial heart has been proved a huge success since December 9, 1982, when the first such implant was installed in the chest of a man. That man, Barney Clark, later succumbed to complications (pneumonia), but the operation was a success and the artificially implanted pumping device worked as planned.

Artificial heart pioneers include Robert Jarvik, whose Jarvik-7™ model was implanted in Barney Clark; Denton Cooley, a Texas physician; William DeVries, the surgeon who conducted four of the world's first five implants; and Willem Kolff, who also invented the artificial kidney (dialysis).

The Jarvik-7™ heart incorporates a pair of pneumatically driven blood pumps. It is fabricated of polyurethane material and aluminum. Special drive lines link the implanted heart to an external power source, the newest variety of which is the Heimes™ portable heart driver, a lightweight device inside a briefcase like box that the wearer can carry with a shoulder strap. This allows the patient greater mobility and freedom than did previous drivers, which were large, wheeled machines.

The artificial heart is nothing less than a marvel. Consider that such a device must beat forty million times per year while reliably sustaining a human circulatory system and you begin to recognize just what a miracle bionics has wrought.

ARTIFICIAL NOSES

Two University of Toronto chemists, Michael Thompson and Ulrich Krull are working on a bionic nose. Equipped with biosensors, the nose works when a fatty compound is squirted through a tiny hole in a specially treated membrane: a sheet of Teflon that is electrically charged. The nose is not yet ready for use, but with additional computer modeling and the use of genetically engineered proteins, Thompson and Krull expect to demonstrate a working model.

ARTIFICIAL LARYNX

At Little Rock researchers at the University of Arkansas for Medical Sciences have invented a valve that can restore speech to more than 60 percent of patients whose larynx, or voice box, has been surgically removed. The valve is inserted in the neck in a surgical incision between the esophagus or feeding tube, and the windpipe. The speech produced sounds almost identical to the speech the individual had before having his or her larynx removed.

ARTIFICIAL JOINTS AND BONES, TENDONS AND LIGAMENTS

Artificial joints in arms, hips, and legs are common. Such joints are usually made of an alloy of aluminum, titanium, and vanadium and often fit into a polyethylene cup or socket. Biotechnologists have also identified and are manufacturing a natural protein substance called *CIF* (cartilage induction factor) that can stimulate the body to repair damaged bone. They may even be able to produce bone in the laboratory. Electrical stimulation is also being used to help knit broken bones. Specialists have devised techniques by which carbon fibers coated with a bioplastic substance become a base upon which scarred, damaged, and torn ligaments and tendons may be healed and regrown. Within nine months of implantation, connective tissue fills the gap in the ligament or tendon and it is once again sturdy and flexible. The three who devised this method are an orthopedic surgeon, a mechanical engineer, and a materials engineer.

ARTIFICIAL VEINS, ARTERIES, AND NERVES

Dacron and polyethylene grafts have been in use as artificial veins and arteries for more than a decade and are now implanted in over a million persons. More than five hundred medical firms and labs in the United States and abroad are working on new types of artificial veins, arteries, nerves, and organs. One such company is Carbomedics, an Austin, Texas, firm that uses carbon-based materials to reconnect severed nerves.

The Biological Alternative: Living Organs and Computers

Today's bionic body parts are made up of either metal, glass, plastic, cloth fibers, or a combination of these. But the science of biotechnology is fast developing the capability to produce artificial body parts and systems that are natural—that is, composed of biological material.

The biochip is almost a thousand times as dense as the most advanced silicon chips-which they will replace. Such biochips are so small that 6,200 of them will he able to fit within the width of a human hair, and a preliminary Gorham report says that, "remarkable near-term applications for biochips include biosensors, artificial intelligence, robot vision, advanced industrial process control, thin film TV tubes, and defense applications."

The biomolecular computer will be a fantastic innovation and create technological possibilities that are almost impossible to envision. Nanotechnology is making this possible. Organic, self-renewing biochips many times as powerful as the silicon chips in today's personal computers and cell phones could drive supercomputers and be implanted in the human body.

In addition to the biochip, new biotechnology methods will actually permit the growth of natural body organs in the lab. Skin is already being grown remotely and then grafted onto burn patients. Collagen Corporation, a Palo Alto, California, firm, has developed a whole series of products that regenerate soft tissue and bone. One product repairs the esophageal sphincter of patients who have suffered severe chronic heartburn; another restores damaged vocal chords.

Two chemical and pharmaceutical industry giants that are working on organ regeneration are Monsanto and Bristol-Meyers. Also, a number of smaller biotechnology firms are actively researching both biochips and organ regeneration.

In the future, nano-sized computers implanted in the human body will scan for disease indicators, diagnose diseases, and control the release of the appropriate drugs.

Bionics and Robotics

The fields of bionics and robotics are inextricably linked; indeed, they are fast becoming indistinguishable. This close relationship was graphically noted recently by a top robotics researcher, Dr. Susan Hackwood, former director of AT&T's robotics program and head of the Robotics Institute at the University of California at Santa Barbara and now executive director of the California Council on Science and Technology. Said Doctor Hackwood, "There are two main goals of robotics that are really different sides of the same coin. One is the goal of building a bio-morphic, or lifelike machine. Many roboticists are, in fact,

DARPA's prosthetic arm give amputees new hope.

Below: Researchers at Vanderbilt University have created this prosthetic leg using smartphone technology.

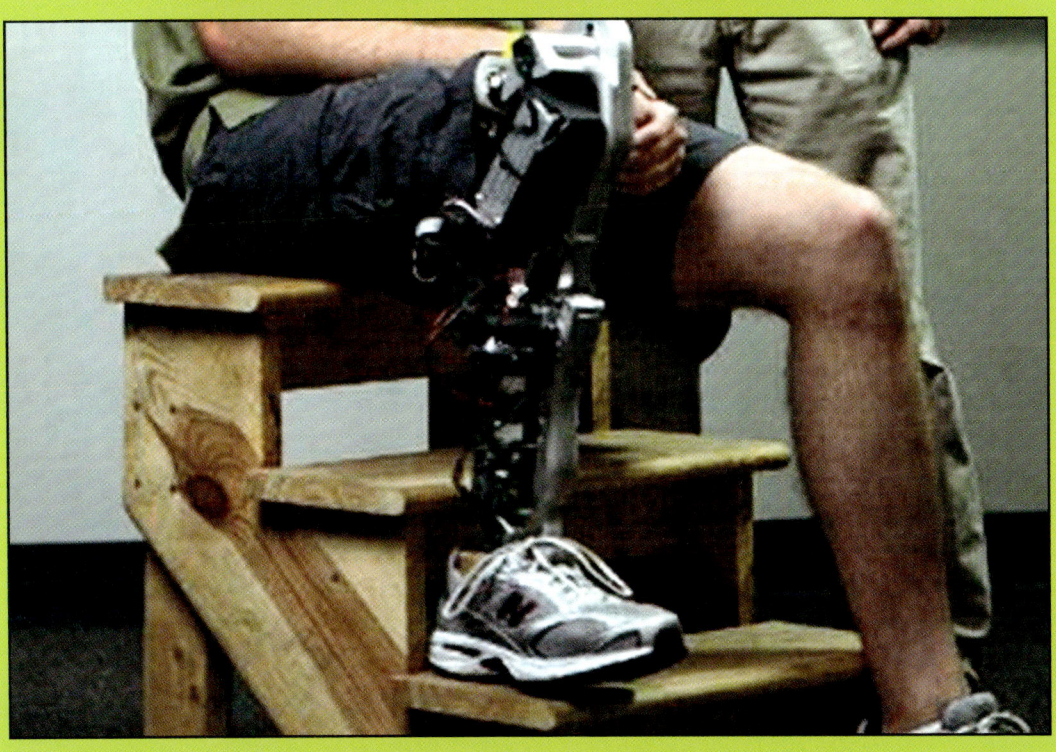

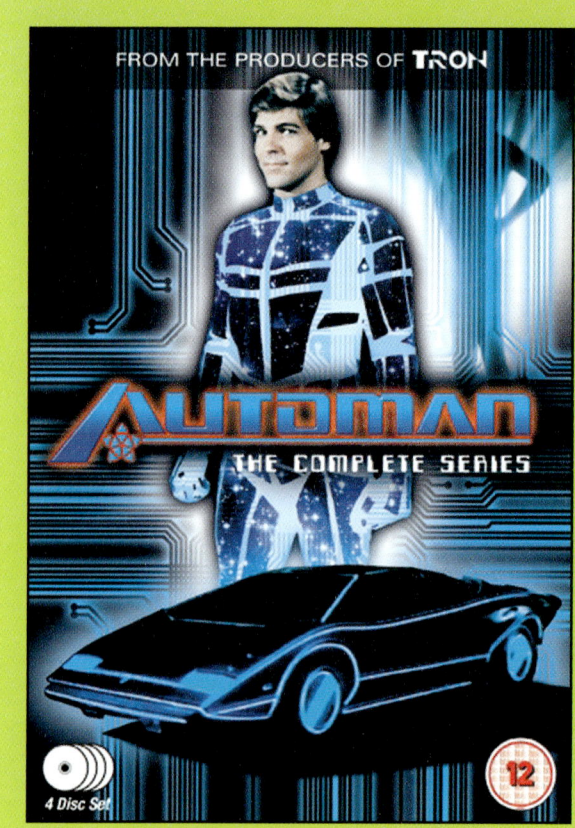
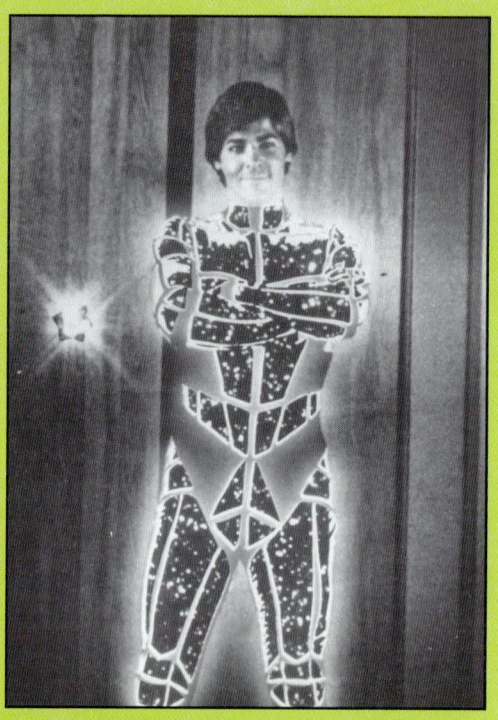

Automan was a TV show.

engaged in building artificial limbs and human prostheses. The other goal is building robots to improve productivity in the factory of the future."

Generally, bionics is most important in its relationship to the study of anthropomorphic robots—androids and humanoids. An increasing number of experts view biological systems and nature as fruitful areas of interest to industry. For example, George Piotrowski, associate professor of mechanical engineering at the University of Florida in Gainesville, contends that the human body is "the ultimate machine." In *Mechanical Engineering* magazine, Piotrowski suggested that industrial engineers apply the biological principles found in the human body to their design strategies. He pointed out that the human brain and nervous system is an excellent example to follow in data processing design, that "the mechanical structure of the body is built of optimally designed links made up of fatigue-resistant material;" and that "human bearings—joints—allow motion with a minimal loss of energy to friction."

Many others agree with Piotrowski. Biophysicist Helmut Tributsch, in his exceptional book, *How Life Learned To Live* (MIT Press) explores the technology of all living things, including humans. Tributsch holds that the human body's biological systems encompass the basic principles of mechanics, thermodynamics, acoustics, locomotion, and optics. He notes that we can find the biological counterpart of cameras in the eye structure of animals and that our most sophisticated ultrasound technology cannot compare with the natural equipment of bats and dolphins.

While Piotrowski, Tributsch, and others tout the human body as a viable example for the design of machines, a growing number of engineers and scientists strive to apply engineering principles that aid the human body. For example, robotics experts and

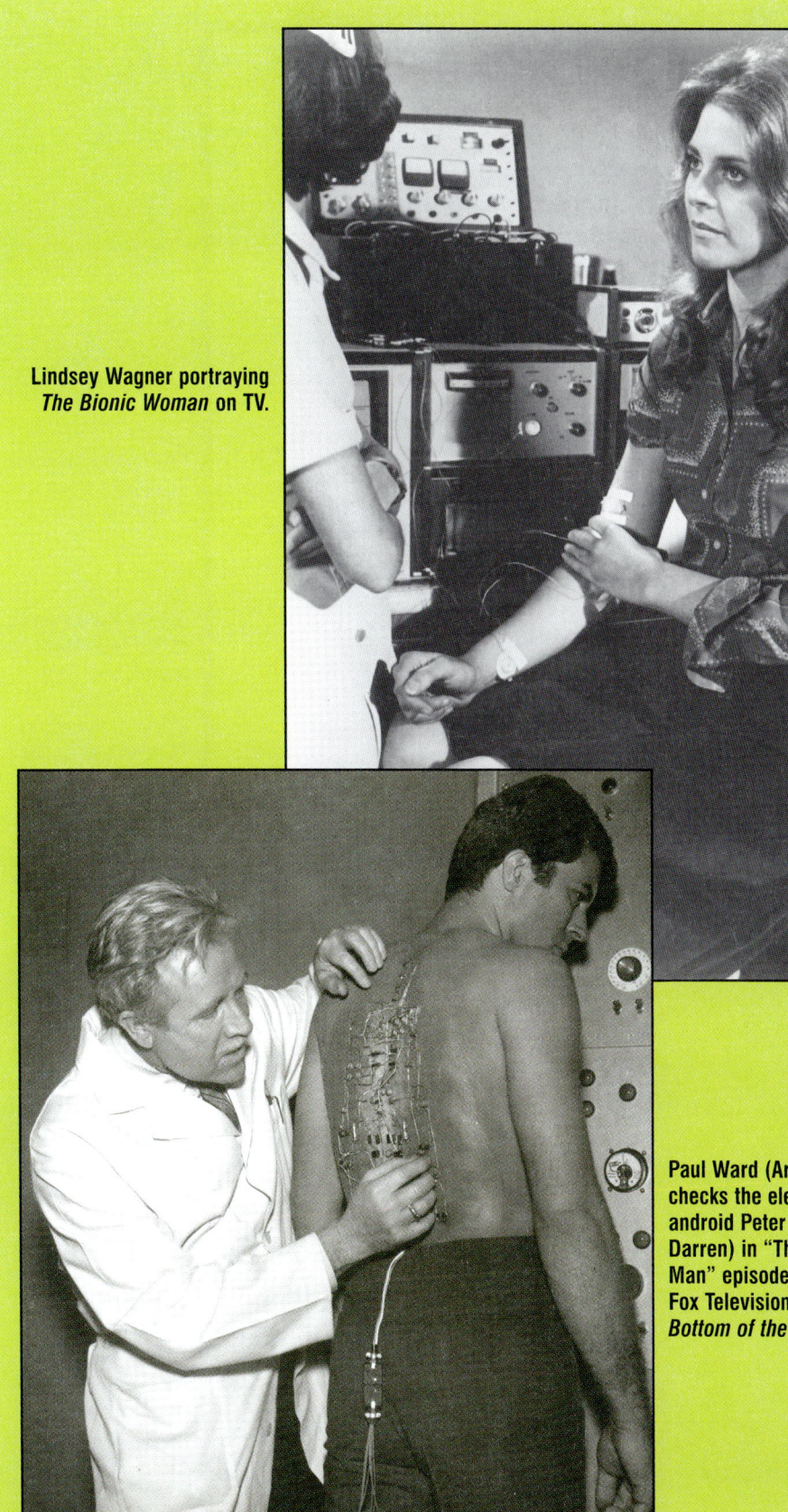

Lindsey Wagner portraying *The Bionic Woman* on TV.

Paul Ward (Arthur O'Connell) checks the electrical circuitry of android Peter Omir (James Darren) in "The Mechanical Man" episode of 20th Century-Fox Television's *Voyage to the Bottom of the Sea* TV series.

Entertainer Christine Aguilera's new album cover.

Below: A laboratory uses these dummies for its autonomous automobile. This give us a brief image of the future world of the bionic human.

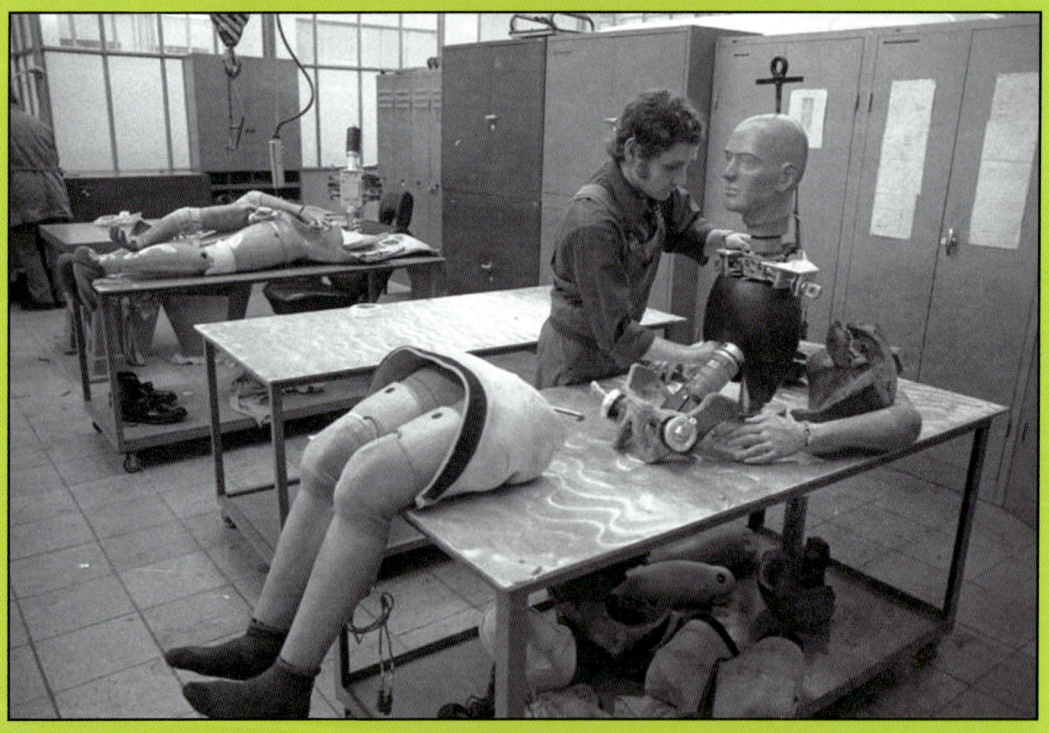

biomedical engineers are developing robotic wheelchairs and arm robots to serve the needs of paraplegics and quadriplegics. Meanwhile, the field of *biomechanics* is an emerging movement oriented toward helping athletes and sports stars achieve optimal physical performance. This is accomplished by studying how the human body can best function as a machine. Video, computer analyses, sensors attached to the body, and other robots are used in biomechanics to measure performance and suggest improvements to enhance hand, arm and leg movement, and body position.

Robots for the Disabled

All individuals need independence, dignity, and the ability to operate in the world as it is. Many physically disabled people are deprived of these needs because they do not enjoy the use of all their body limbs and organs. Robotics and bionics research is producing a number of systems designed to assist them. Robots have been built to assist the disabled in daily activities and to help restore the ability to successfully interact with the world.

Currently, a number of robot devices are available. One robot device permits handicapped persons to telephone without using their hands. Basic Telecommunications Corporation offers the Ability Phone, which consists of a visual display, a keyboard and an optional voice synthesizer. A blind or arms/hands impaired person can command the phone to dial a certain number. With the keyboard, a mute person can type a message that will be transmitted by voice to a distant party. At Stanford University, Dr. Larry Leifor has tested a robot that answers the telephone, operates a typewriter, and turns book pages.

Computerized robots are providing fresh hope to the handicapped. Carl Mason, head of research at the Veterans Administration Center in New York state has demonstrated one machine that can shave an armless man or one whose arms or hands are infirm. The man directs the robot through spoken words and moves his face to improve the shave. The New York center also is experimenting with working robots that assist disabled persons to eat, dress, and groom themselves.

Robotic wheelchairs are also proving to be a boon to the disabled. One model developed by the Veterans Administration has a built-in robot arm. Using chin pressure, a person whose hands are paralyzed can grasp objects and can even play games like chess. Another robotic wheelchair is computerized and permits the quadriplegic to blow into or sip on a straw to command the wheelchair to go forward or move in reverse.

In Dayton, Ohio, researchers demonstrated that a paralyzed paraplegic could use his/her own body as a sort of "robot platform." A brave young woman named Nan Davis, a 22-year-old paraplegic and a senior at Wright State University, was seen by millions on network TV news programs, as she stood and took a half-dozen faltering steps. This miracle came about as a result of a microcomptuerized control system developed by Dr. Jerold Petrofsky. An array of electrodes and sensors placed on Nan's legs, knees, and ankles fed information into a remote computer that sent back timed impulses to the limbs to move. Said a smiling Ms. Davis after she had completed her short but breathtaking jaunt, "One step for mankind."

The Future

The host of complicated gadgetry used to help Nan Davis take those first miraculous steps is crude compared to the devices that will be available to implant miniaturized control devices and prostheses in the body of a crippled person to serve as an artificial spinal cord and nervous system. These devices may not even be observable as the individual moves about in society.

Organ regeneration will also provide medical specialists with a continued supply of body parts, and biosensors made of natural, bio-technologically produced materials will assist bodily functions when appropriate. Joseph Moskal of the National Institute of Health says that biochip devices will soon miraculously provide sight to many blind persons and help paraplegics regain control of their muscles. Because of the minute size of the chips, Moskal adds, they "could be implanted anywhere in the body."

The wonderful possibility is that disabilities will be healed, lives will be saved, and wheelchairs may become a thing of the past—a relic of early science and medicine. In effect, the fields of robotics, bionics, and biomedical engineering will merge and once handicapped individuals may become our first cyborgs.

It also seems likely that continual progress in biotechnology, bionics, and robotics will serve to provide humankind with robot systems that are made up of the best biological, mechanical, and electronic components, processes, and structures. The likelihood is that if research and discoveries continue at their current rapid pace, we will end up with vastly improved robots capable of independent thought and flexible motion. Simultaneously, medical technology will make possible the replaceable person, with hundreds of body parts available on the shelf.

Chapter 11

"Your Slippers, Sire"—Personal and Home Robots At Your Service

Who hasn't dreamed of someday owning his or her very own personal robot: a friendly pal and servant who can do all those onerous household chores. Most of us would want our home robot to be capable of washing and putting away the clean dishes and taking out the garbage. Then, our home robot would vacuum the carpet, and make the beds. And when we come home from the office or business place after a hard day's work, we would want our robot to greet us at the door, serve us a drink, and bring us our iPad or tablet as we relax on the sofa. Later, after we've relaxed, we might even appreciate some light conversation from our witty, charming, and uncomplaining electromechanical companion.

If all this is your dream of what a personal and home robot should be able to do, then the models now available may well disappoint you. Someday, a little later on in the twenty-first century, robots will be able to do all these things. It might even be your intellectual equal or superior, challenge you to a game or contest, maybe even engage in a romantic scene with you.

For now, the personal robot can come home with you, but once there, it will not perform anywhere near the level of R2D2, C-3PO, or some of those science fiction robots you may have read about.

On the other hand, if you are willing to lower your sights just a bit and accept today's personal robots as the technological marvels they are—limited in ability but marvels nonetheless—you might wish to seriously consider acquiring one of the latest models. Putting things in perspective, the fact is that ours is the first generation to actually have the wondrous opportunity to buy a robot and take it home.

Most experts acknowledge that the available personal and home robots are not yet capable of being all-around servants and companions, but they emphasize that the machines are indeed wonderful educational tools and fantastic entertainers. And you may just be surprised at the capabilities. Mobile (on wheels), they talk (using speech synthesis devices), play games, transport items around, and crack jokes. A few sing songs. Several can dance by rocking to and fro. Many a robot can be operated by voice command, responding to its name when called by someone whose voice it recognizes.

As we have seen in earlier chapters, more expensive robots have on-board computers,

Servitron's able robots help to entertain guests. Inflatable, the robots are radio controlled. Made by Serviton Robots of Denver, Colorado.

with a keyboard for data entry and programming. Several companies offer software that can enable computerized home robots to act as burglar alarms and as fire detectors. And several personal robots have personalities. Robots from The Robot Shop, a California company, show emotion. They can play blackjack or other simple games with humans. When they lose, they throw a temper tantrum, whining and shaking their body furiously just as would a human sore loser. However, if the robot wins, it lets out a shriek of joy, wobbles back and forth, and lets everyone know how happy it is.

How much your personal robot may cost depends on how sophisticated your expect the robot to be. Robots with on-board computers and packed with electronic sensors may set you back anywhere from $5,000 up—a few models cost as much as $20,000. You can have a deluxe robot custom-made to your specifications for as much as $100,000! On the low end of the cost spectrum, you'll find dandy robots like Droidbug and similar turtle-like models that retail for less than $100. These bargain-priced robots may not do a lot, but as entertainers they can provide hours of enjoyment and fun, as well as serve as educational devices.

Robot Showcase

To acquaint you with some of the personal home robots, here's a showcase of commercial robots, picturing many of the models and describing their capabilities. Many of these, regrettably, are no longer available. For more information about personal and home robots, including manufacturer addresses, price list, specifications, and performance characteristics, we recommend you google up "personal or home robots."

HERO 1

The popular Hero 1 robot opens the show. About 25,000 Hero 1s were sold. The manufacturer was Heath Company, whose Heathkit division has a well-deserved reputation for quality electronic products.

Hero 1 is billed as an educational system. The robot comes complete with textbooks and educational software to help you learn about robotics. This isn't a snap robotics course so you should expect to spend about 125 hours in independent study completing the Hero 1 fundamentals curriculum. Topics you'll learn include electronics, mechanics, and computer basics.

Hero 1 is computerized and is programmable through a small hexadecimal (six-key) keyboard. However, since it is difficult to program Hero 1 to perform even the most basic of tasks, don't expect too much in the way of performance without being willing to spend the time necessary to program the robot. This is true not only for Hero 1 but for most other programmable robots.

This capable robot is packed with interesting features. Its single arm rotates 350 degrees, as does its head. The head contains sensors for a motion detector, light detector, ultrasonic ranging system, and sound detector. A four-year calendar clock is included. Hero can also speak and make sounds with a robotic accent.

HERO, JR.

Whereas Hero 1 is designed to be an educator, his chummy little offspring, Hero, Jr., is strictly a home machine. Because Hero, Jr. is already preprogrammed, you don't need any specialized knowledge to operate this robot.

Preprogrammed to speak, sing songs, recite poems and nursery rhymes, and play games ("Cowboys and Robots"), Hero, Jr. also acts as a security guard, alarm clock, and appointments clerk. He can be operated by wireless remote control. The robot likes people. With his infrared detector, he seeks people and moves toward them. Hero, Jr. is a fun robot with a built-in personality. "He's sentimental, sophisticated, and at times a real ham," says Wayne Wilson, product manager at Heath Company.

RB5X

In September 1982, RB5X became the first mass-produced personal robot to be sold commercially. This small fella (23-inches tall), who resembles R2D2 of *Star Wars*, is a very advanced creation. RB5X is perfect for people who want to learn about robotics and electronics, because they can order a companion set of textbooks and workbooks. Many schools are adopting RB5X (and Hero 1) for their classrooms. RB5X is also being sold in Japan and Germany—in Germany, he's called "Toby."

RB5X is a fine family companion. Computerized, he possesses sonar and tactile

Production line for Hero, Jr., manufactured by Heath Company. Sorry, he's no longer available.

Hero 2000 demonstrates its useful arm. This robot was a big hit of Heath Company, which also produced Hero, Jr.

The Heath Company offered a complete, hands-on robotics and industrial electronics course.

ROBOT ALCHEMY • 157

I have the slippers. Could you get the paper?

In the 1980s, the RB Robot Corporation ran this interesting advertisement in magazines like *Creative Computing* and *Popular Computing*.

RB5X shows us his "motherboard" and inner parts. (RB Robot Corporation)

sensors and a wide array of options, including a fire extinguisher attachment. There's also a vacuum cleaner attachment that allows the robot to vacuum your living room—after you have left for work. RB5X has an arm for lifting and carrying light objects (like slippers), and he can speak.

HUBOT

When Hubot first came out, iin 1985, Michael Forino, president of Hubotics, said that his company's robot is "the ultimate home appliance." Forino backed up his assertion by pointing out that this sophisticated robot was a computer, a television, a stereo cassette player, and a robot all rolled into one.

The formidable Hubot easily moves around the house, avoiding obstacles by using his sonar collar to detect them. He serves guests with his tray while chatting pleasantly. He comes with a built-in computer and attached keyboard. Because he has a 12-inch TV screen (which displays his face), Hubot can recite your personal schedule for the day by displaying it. He also is a delightful partner with whom to play educational games, including Atari 2600 video games, and he acts as a burglar alarm and fire alarm system.

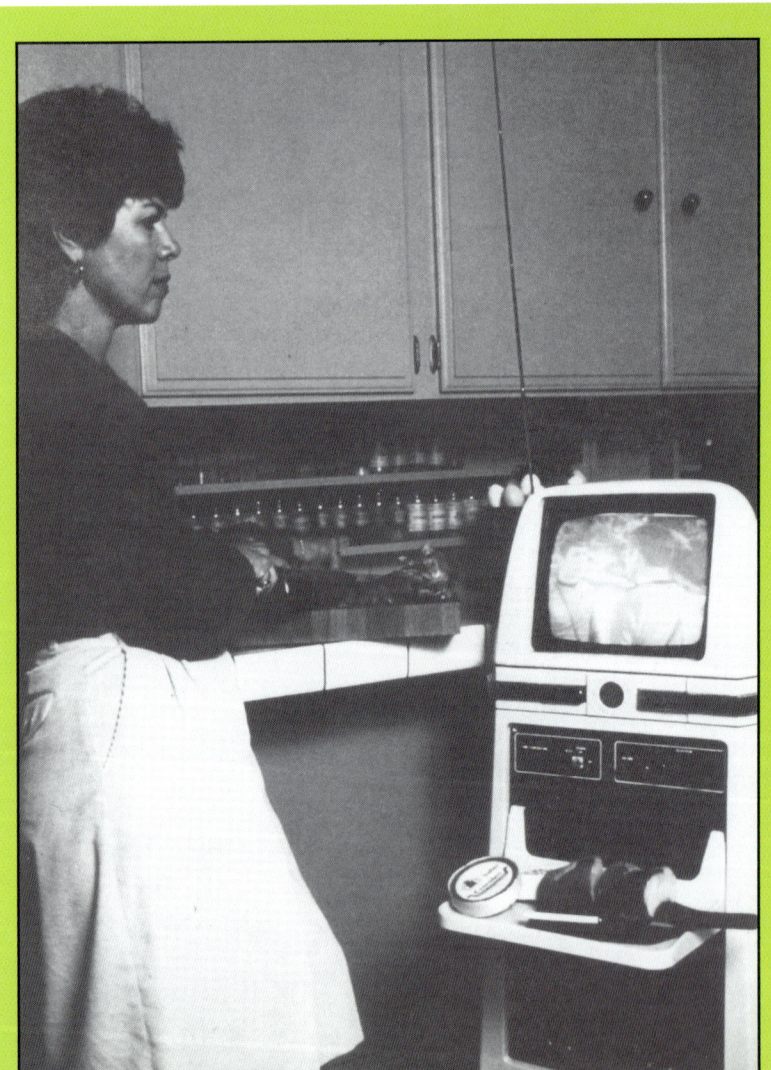

Hubot, the robot, served as a great kitchen aide and director.

Tomy Toy Co. came out in 1984 with its lineup of the lovable Dingbot (left), Omnibot (middle), and Verbot, a voice activated model (right).

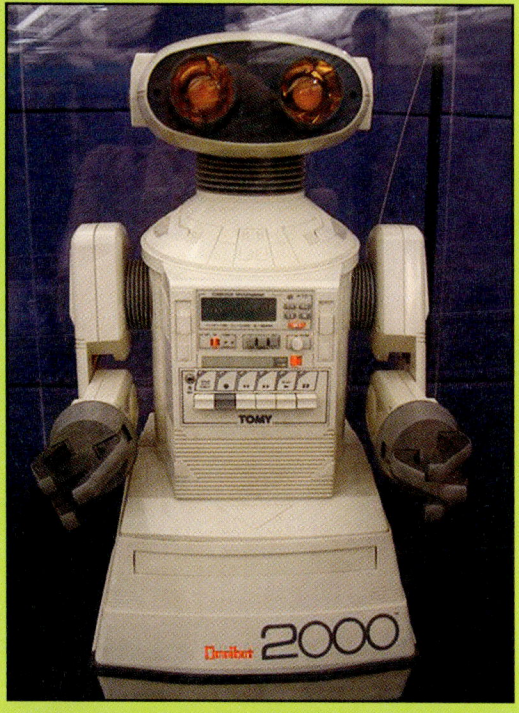

In 2000, Tomy Co. introduced Omnibot 2000. Later, the company came out with a version known as i-Sobot.

Omnibot, (middle) the pre-programmable electronic robot, helps to cater the party.
Verbot, (above) a voice-activated robot, responds only to your voice and secret commands.

Comro Tot brings drinks to guests (left) and visits with a seal at Sea World. (Comro Tot Co.)

Weighing about 100 pounds, Hubot is 44-inches tall. Sorry to everyone who wants one, Hubot is no longer available to order.

OMNIBOT

Here is a brainy robot that doesn't cost a bundle (priced at less than $300). Made by Tomy, a large international toy company, Omnibot is a preprogrammed home robot operated by a remote control unit. Responding to your commands, this tiny machine person can deliver a poem or walk on a tour of your house. His memory stores as many as seven programs, which he can perform automatically any time and day you specify in advance.

When Omnibot tells the kids a ghost story, everyone listens because the robot delivers his stories with strobe lights eerily flashing. Omnibot also has an on-board clock with alarm.

COMRO TOT

Comro Tot first gained fame in 1983 when a major cable movie channel decided to use him as a promotional device. Comro Tot was offered as first prize in the "Great Robot Giveaway Sweepstakes," a contest that drew more than 50,000 entries. An early version of Comro Tot was offered in the fabulous Neiman-Marcus catalogue.

Comro Tot is a fully programmable, mobile computer, with a hexadecimal keypad. The robot can carry out a number of household tasks such as serving refreshments and sweeping the carpet. He talks, using his "type and talk" speech synthesizer, and has an ultrasonic security system to catch would-be intruders. When he sees a burglar, Comro Tot shouts, "I see you... You are under surveillance!"

GEMINI

Carl Helmers, publisher of *Robotics Age* magazine (now no longer in print), said he was high on Gemini. This robot, Helmers says, "is a significant step forward in the practical design of autonomous robots for research, personal use and experimentation."

Of course, that was some years ago, and Gemini was, indeed, an advanced home robot. For one thing, it can go through doors and navigate difficult paths by its employment of infrared room beacons and passive infrared reflectors. Gemini is powerful: the robot has two motors and three 12-volt batteries. It can talk (256 words) and obey spoken commands.

The size of a human, Gemini is controlled by three—yes, three—built-in computers. Turn Gemini on and the robot automatically checks its circuits and gives you a status report. There is also an on-board smoke detector, a full-function keyboard, a barometer, and a security alarm system.

ELAMI, JR.

Elami, Jr. is a Canadian product sold exclusively in the United States. Elami robots come

The Gemini

Above, Elami, Jr. says, "Hello, I am Elami, Jr., your electronic friend." Elami, Jr. is sold now by Robotland.

Above: The author's grandson, Brian, has a smile on his face as he is surrounded by Maxx Steele and other toy robots in a local department store.

Right: Topo transports a young girl's school supplies in his optional androwagon.

Below: F.R.E.D. (Friendly Robotic Educational Device), an Androbot, Inc. product, carries a pen with him.

in three models, each of a differing size and level of sophistication. Each robot has a computer, a 2-color Liquid Crystal Display face, a light display and two motorized arms and grippers. An infrared sensor allows the robots to detect and avoid obstacles. You can buy the Elami, Jr. robots in any of three heights: 12 inches, 18 inches, and 36 inches. The larger model has a 9-inch color monitor face and is more powerful physically. It also has a 2,000-word vocabulary.

TOPO

Topo III was built by Androbot, no longer in business. The company was founded by the dynamic Nolan Bushnell. Androbot also made Androman and F.R.E.D. Topo was touted as "a personal extension of your home computer." Topo can talk, sing ("Daisy, Daisy" and other tunes), dance, spin around, and is a captivating entertainer. Poor Topo is armless, but he does have extras like heat and sound sensors. Topo comes equipped with two computer microprocessors.

MAXX STEELE™

Maxx Steele™ is an inexpensive, mass-produced robot that was sold in major department stores and toy outlets. Maxx Steele is the trademark of the Ideal Toy Company, which has created Maxx as the leader of its Robo Force warrior robot series so popular with youngsters.

Maxx comes in two versions: as a fully assembled robot or in remote control kit format. The fully assembled version is programmable, is about two-feet tall, has a moving arm and possesses a 150-word vocabulary. The erection kit model allows its owner to learn some basics of robot construction by way of the assembly experience. Only the assembled Maxx Steele has speech capability.

F.R.E.D. AND ANDY

F.R.E.D. (Friendly Robot Educational device) and Andy are robot lookalikes. Their heads resemble Topo III's. However, each of these two robots have decidedly different capabilities and talents.

F.R.E.D., made by Androbot, can be your pen pal. If you create patterns and drawings on your computer screen, this mobile robot will draw corresponding designs on a sheet of paper or poster board. F.R.E.D. can also speak and avoid obstacles.

Axlon Corporation's Andy can't draw, but his personality is more humanlike than F.R.E.D.'s. His eyes blink, and you can change Andy's mood by adjusting his tone generator. A low monotone makes Andy act sad while a sprightly sliding pitch makes Andy appear quite happy. Andy comes equipped with a light and sound sensor. Software is included in the basic price.

ANDROMAN

Androbot's AndroMan was only 12 inches tall, but he had special characteristics. He was a real-life video game. With an Atari 2600 video computer and AndroMan's game cartridge, you could control this exciting robot with a joystick, using the built-in infrared data link. AndroMan could talk, and he was sort of a cheerleader, prodding you along as you play the video game.

AndroMan was a neat game robot activated with a joystick.

ARMADILLO

The Armadillo robot is not just a fun device, it is a positive aid to education, taking programming out into the real, three-dimensional world instead of the flat two-dimensional world of the video display unit. When connected to your home computer, Armadillo runs around under computer control, moving forward, backward, right, and left with independent control of each wheel. It has blinking eyes, will bleep with a choice of two tones, and, when ordered by the computer, presses down a pen to chart its progress and provides a hard copy of the results of the program. When set free to run around, Armadillo investigates its environment. Should the robot shell bump into an unmovable obstacle, touch sensors send back data to the computer for it to calculate evasive or exploratory action.

WHEE ME

The Robot Shop (robotshop.com) accurately bills itself as "the world's leading source for personal and professional robot technology." It offers a number of unique robot kits for professionals and hobbyists. Many of the computer-controlled robots have unusual sensory capabilities. Some shiver if the room temperature gets too low, and turn red when they become confused. Other capabilities include smoke detection, light detection, solar-charged batteries, night-light, guard duty package, and ultrasonic range finding. You can configure your Robot Shop robot to make funny noises, blink its eyes, light up its heart, squirt water from a water gun, and inflate balloons. You can order the *Whee Me*, a palm-sized massage robot, the Neato VX21 vacuum cleaner, or much larger models. You will enjoy viewing this firm's website.

Tomy's Armatron was both a toy and an educational device. It was distributed by Radio Shack.

Two completed Robot Shop robots that were assembled from kits.

Axlon's Talkabot was a great and fun toy for kids.

Compubot, by Robot Shop

Memo Robot came in a handy transport vehicle.

Chapter 12

Industrial and Worker Robots

It was in 1961 on a General Motors assembly line that entrepreneur/engineer Joseph F. Engelberger installed the first industrial robot. Engelberger, now known as the "Father of Industrial Robotics," had founded his company, Unimation, a few years earlier to the amusement of fellow engineers, few of whom believed the robot could ever be more than a toy.

Today, Unimation is a unit of its parent corporation, Westinghouse, and its robots are hard at work throughout the world. In addition, Unimation now has lots of competition. Hundreds of other manufacturing firms, including General Electric, Sony, and Toyota, are producing robots.

The industrial robot revolution has taken a long time in coming. Japan's largest corporations were first to recognize the tremendous advantages of robotics. As a result, that country invested heavily in worker robots and, as we have seen, currently is the

Isaac Asimov (left), famous robotics science fiction author, with industrial robotics pioneer, Joseph Engelberger.

Joseph Engelberger, often call the "Father of Robotics," installed the first industrial robot on a General Motors assembly line.

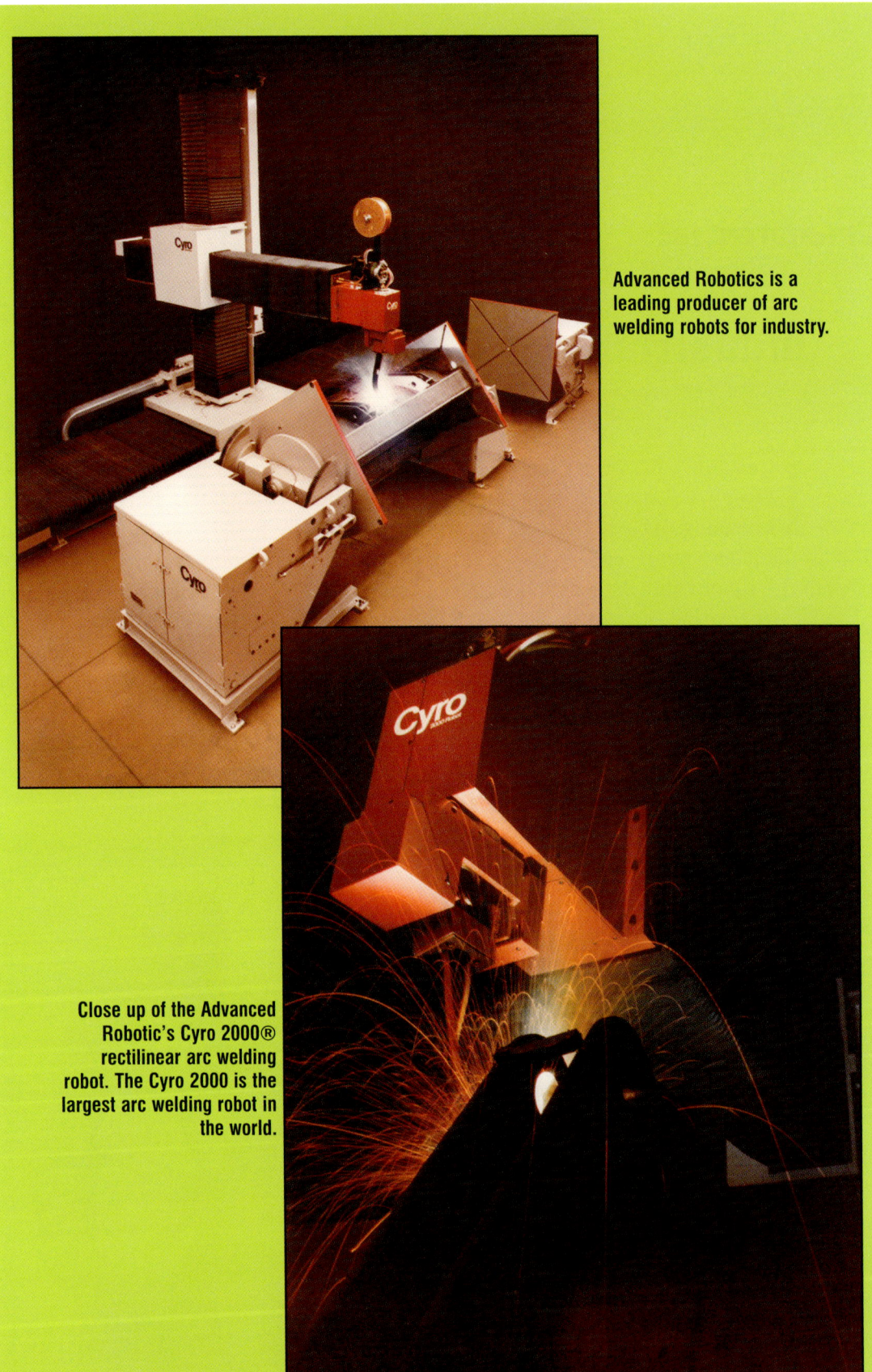

Advanced Robotics is a leading producer of arc welding robots for industry.

Close up of the Advanced Robotic's Cyro 2000® rectilinear arc welding robot. The Cyro 2000 is the largest arc welding robot in the world.

world leader in robot employment. In America and Europe most manufacturing corporations were reluctant to buy robots and automation equipment. However, in 1980, there was a rush to automate the United States economy, and robot sales took off.

The Plain Jane Worker Robot

Industrial robots aren't as "pretty" or as cute as the household variety of robot. They're usually huge, one-armed, steel-muscled creatures with a remote or on-board computer serving as micro-brain. But pretty or not, their value is great in terms of labor costs saved, and their genuine utility has been well demonstrated. On TV's "Ripley's Believe It Or Not," two industrial robots showed their dexterity by making a ham and cheese sandwich!

Industrial robots are functionally configured; that is, they are built to perform specific tasks. Therefore, it isn't essential or even desirable that

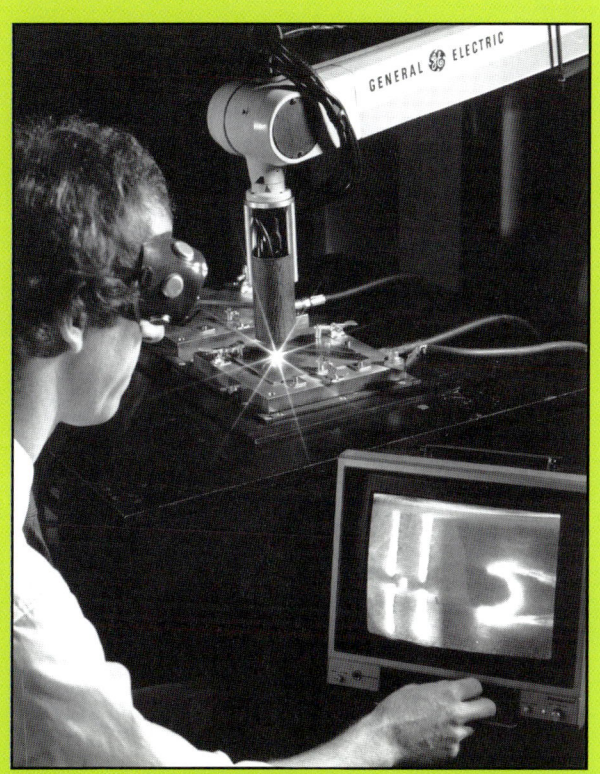

This General Electric robot has a sensor system that brings the powers of sight and intelligence to welding robots. The advanced vision and control system enables a welding robot to steer itself along irregularly shaped joints, continually observing the joint and weld "puddle" and making adjustments as it travels along.

they be shaped in human form, or that they possess humanlike senses. Some industrial robots, for example, have changeable end-effectors, or hands. GM Fanuc Robotics offers automatic handchangers that enable an operator to change robot hands and tools in only two seconds, permitting the same robot to perform several different operations at a single workstation.

Why Robots?

Robots undeniably increase efficiency and cut manufacturing costs. Studies have shown that an industrial robot, though initially expensive, can pay for itself over a one to three-year period. One robot can replace as many as ten human workers, depending on the specific job or task.

Robots are more quality conscious than human workers. Once programmed properly and put to work, they do the job right—first time and every time. Manufacturers with newly installed robots report significant savings in materials. In addition, the best robots average 98 percent up-time and work at peak efficiency hour after hour, day after day, three shifts per day.

Says an expert at ASEA, a large robotics firm, "Robots free human workers from dull, dangerous, dirty, and demeaning tasks, permitting [them to take] jobs for which they are best suited."

A single robot can replace up to 10 human workers.

Automobile Workers

Most industrial robots work in automobile plants. At Chrysler Corporation, the number of robots has increased more than 50 percent each year for a decade. In 1975, the company employed only 16 robots; ten years later that number had grown to 570. In 1988, Chrysler had 1,500 robots and today, most of the work is done by robots. The same startling statistics can be found when we study the robotics experience at General Motors and Ford in the United States and at plants in Europe and Japan producing Fiats, BMWs, Mercedes, Toyotas, Hondas, and other automobiles.

The work accomplished by robots at auto assembly plants is varied. Robots weld, solder, mount brackets, torque bolts, mount wheels, paint bodies, seal windows, die cast, and move parts along assembly liens. They're also used to shift gears in tests of manual transmissions and are increasingly being upgraded to accomplish additional tasks. The newest generation of robots is re-programmable, permitting rapid retooling and line operation adjustments. Richard Johnson, Chrysler's engineering director, says that almost no human jobs are sacred and robots are even used to carefully shave and shape clay models of tomorrow's cars. This design work is computer-aided.

Computer-control of robots greatly enhances their usefulness. New auto plants planned or under construction employ sophisticated computer-aided design/computer-aided manufacturing (CAD/CAM) systems and robots. This combination of resources, known as *computer-integrated-manufacturing* (CIM) promises to revolutionize the auto industry and almost every other manufacturing industry. CIM will give us the Factory of the Future: a place where automation is complete and humans are few. It seems likely

Robots on the auto assembly line.

Modern robot packers at work packing bread.

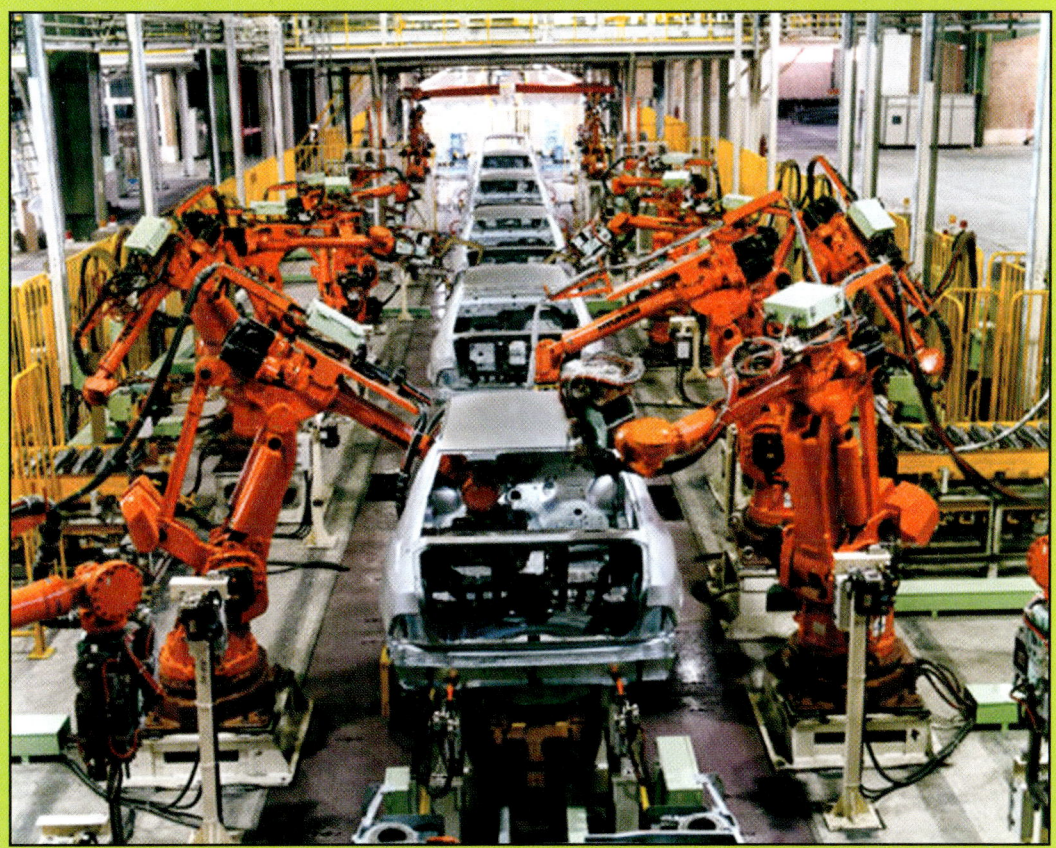

Robots have replaced millions of auto workers around the world.

A robot painting pin-stripes on an auto.

that as the twenty-first century progresses, robots will replace most human workers in factories. Proponents of technology say that humans will find other jobs; that, in fact, robotics will spur the economy and produce additional employment. But more astute observers of 21st century robotics and automation aren't so sure. Some predict widespread unemployment when CIM takes firm hold in the world's economy.

What is sure is that robots are becoming more intelligent and infinitely more valuable. Charles Rosen, chief scientist at Machine Intelligence Corporation, notes that yesterday's robots were "stupid." Some lunged, grabbed, and jerked in irregular motions. But, adds, Rosen, the new generation of robots is vastly improved: their actions are smoother. Phillip Villers, a robotics pioneer now with Automatix agrees. "The robot's muscle has given way to more intelligence and finesse," Villers says.

Worker Robots Multiply

Once exclusively in auto plants and large factories, robots have more recently begun to move into other areas of the economy. We now have robots in offices, on the farm, in homes, and in classrooms, along with an expansion of robotics applications in industries such as chemicals, baking, shoe-making, textiles, and oil drilling.

In addition to huge industrial robot machines, light assembly robots are beginning to find uses in small manufacturing shops and even in restaurants. These smaller robots—many of desktop size—can assemble electronics products, dunk small parts in chemicals, and inspect parts for seam cracks and defects. Fast food restaurants, banks, space, the armed forces, police and fire departments, nuclear plants, book stores, printing plants, underwater, and in the entertainment industry. You might be surprised at some of the unusual jobs now being performed by these workers.

Fifteen Unexpected and Unusual Jobs for Robots

1. *Railroad Laborer*. In Virginia, the Norfolk and Southern Railroad uses seeing robots to handle and inspect railroad ties.

2. *Waiter/Waitress*. The Two Panda Deli, a Chinese restaurant in Pasadena, has a real treat in store for diners. Two robots, Tenbo R-1 and his girlfriend Tenbo R-2, are

employed as waiters. This duet delivers food, plays music, and takes orders from customers. They don't always understand instructions, however, and when they get confused, the robots exclaim, "Do not understand" or "That's not my problem." The diners enjoy the robots, especially when a new song comes on and the pair scream out "Get down and boogie!"

3. *Fast Food Restaurant Jobs*. In Baltimore, a robot company is building an experimental robot for a large hamburger chain. The robot will have six arms, cook and serve food, take orders, and count money. It will cost $100,000 but might replace a dozen human workers. If it is vandalized, the robot will flash an alarm to police.

4. *Bartender*. Scarab Robotics Corporation is marketing a $65,000 robot bar. Called the Scarab X-1, the robot bartender can take orders placed by remote voice command or through a keyboard. It can make four drinks a minute, total up drink costs, verify credit cards, and communicate with the cash register.

5. *Flight Instructor*. Realistic robot flight simulators are now training military pilots. The human pilots get to practice their knowledge and learn fresh procedures as the simulator puts them through astounding flight situations. The robot simulators look and feel exactly like a real aircraft cockpit.

6. *Nuclear Plant Inspectors*. Many nuclear plants are bringing in robots to perform the

This Japanese robot is excellent to reach into those difficult places.

Designed by Rodney Brooks, the "Baxter" humanoid robot can be programmed to perform many simple jobs never before robotized. it is perfect for a smaller factory. "Baxter" is made by Rethink Robotics of Boston, Massachusetts.

dangerous task of inspecting areas for radiation leaks, and to work in places where leaks have occurred. At the Three Mile Island nuclear plant in Pennsylvania, where the nation's worst nuclear plant accident occurred in 1979, a robot called "Rover" recently was the first to be lowered into a basement area where there was radiation. "Rover" reported that the radiation inside the area was twice as high as a human could safely stand. At Fukishima, Japan, robots were also used.

7. *Telephone Lineman*. Two robot companies are jointly building robots to climb telephone poles and high voltage power towers. A human lineman on the ground will be needed to supervise the robots' work.

8. *Medical Aides*. Japan is building robot nurses strong enough to lift patients. The nurses will also deliver medicine in exactly the right amounts to patients on hospital wards.

9. *Window Washer*. Do robots do windows? After the city fathers had built the Tower of the Americas for San Antonio's HemisFair in 1968, a big problem came to their attention. Who was going to take the risk of washing the windows of the revolving restaurant and observation deck—750 feet above the ground! To the rescue came a

robot window washer manufactured by Southwest Research Institute. The machine, controlled by technicians from inside, has been doing a magnificent job ever since.

10. *Fire Fighter*. When the Phoenix Fire Department gets a distress call that someone is trapped in a burning house, they call on Snail, a mobile robot fire fighter. Snail has a fire hose attachment and can venture inside flaming homes and buildings with structures in danger of collapse. Another fire fighter is Odex I, the amazingly powerful "functionoid" machine built by California's Odetics, Inc. Odex I can fight home and industry fires, rescue passengers trapped inside burning aircraft, and battle raging forest fires. If something heavy falls and pins a human, the robot can easily lift it off: he's able to carry loads up to 1800 pounds.

11. *Bomb Disposal Technician*. Many cities use robots for bomb disposal. When New York City's Mayor learned that three courageous police officers had been seriously injured handling a terrorist bomb, he ordered the city's police department to immediately purchase three remote-controlled robot bomb carriers.

12. *Supermarket Assistant*. The butcher, the baker, and soon the candlestick maker may be robots. In a Nokendai, Japan, supermarket customers are greeted by robots that announce special sales. In the meat department, a robot butcher slices the meat order to customer specifications. In the stockroom, robot carts deliver merchandise to the proper shelves, and the cash registers are laser- and computer-controlled. Next, the supermarket's owners plan to put robots to work bagging groceries and taking them to customers' cars.

13. *Underseas Explorer*. Robot submersible craft are incredible performers. In 1981, the National Geographic Society used a 1,450-pound underwater robot to explore and

This underwater robot can clean up trash and other items from a canal or other water source.

photograph the wreck of the British ship *HMS Breadalbane*, a vessel that sank in the icy depths of the Atlantic Ocean in 1853. Unmanned robots at Woods Hole Oceanographic Institute, in Massachusetts, are now being used to dive 20,000 feet, mapping ocean floors for a month at a time. The U.S. Navy uses robot undersea craft for a variety of missions, some of them classified top secret.

14. *Surgeon*. Would you trust a robot to perform surgery on *your* brain? Experts believe the precision accuracy of robots may prove to be preferable to the sometimes uncertain dexterity of a human surgeon's hands. The first brain operation in which a robot was employed was in 1985, at Long Beach, California's Memorial Medical Center. A Unimate PUMA robot arm controlled by computer performed a needle biopsy on a 52-year-old man with multiple brain tumors. The patient said he had "no qualms" about the operation. Conscious throughout the operation, the man left the hospital three days later. Today, there are many such medical operations performed by robots.

15. *Greenhouse Worker*. "Green thumb" robots are on the way. *Robot/X News*, a robot industry publication, reported that several companies are developing robots that hang from overhead gantries in greenhouses. Computer-controlled, the robots apply pesticides, water, and fertilizer at prescribed levels.

Chapter 13

Entertainment Robots

One type of robot worker that also deserves and attracts our attention is the robot promoter and entertainer whose specialty is enticing customers. Radio-controlled robots are great at publicizing all kinds of products.

Robots in a variety of shapes and sizes are being used for promotional stunts and hoopla. One in the shape of a Coke can is used by the Coca-Cola Company; another, configured as a milk carton, makes the rounds at Borden's Dairies. Robots are wonderful entertainers, and their appearance for events such as a trade show, computer fair, or at the toy department of a department store guarantees a large crowd and lots of publicity.

Show robots aren't limited, however, solely to the promotion of products. Take for example, the work of DC-2, a show and promotional robot created by Gene Beley of Android Amusement, an Irwindale, California, robot company. DC-2 has become famous for his exploits. One man hired the robot for a morning to protest California's divorce laws. Roaming the sidewalks near the Redwood, California, courthouse, the robot blared an appropriate country song, "She Got the Gold Mine, I Got the Shaft!" The media, attracted by the large crowds that gathered, carried both TV and newspaper reports of the incident.

Mr. Ritz Miller, the protestor who had hired DC-2 for $1,000, declared the money well-spent.

Walking, talking demo robots by Heinz bring fun and zing to an exhibit or show.

At Anne Arundel Community College in Maryland in 1984, a "distinguished" robot was the guest speaker for commencement exercises. His topic, naturally, was "the future of high technology." Here the robot is interviewed by a local television reporter.

"DC-2 did a better job for me," Miller explained, "than the attorneys and courts to whom I gave over $65,000."

Of course, DC-2 can perform many other tasks, too. He appears regularly on TV shows, and actor James Caan presented a DC-2 look-alike robot to a publisher friend as a Christmas gift. Caan's friend uses the robot as a household helper to greet guests, serve drinks, and dance at parties—all tasks easily handled by DC-2, who has a built-in serving tray and loudspeakers.

Touring the Universe of Show Robots

Every day, robots are performing somewhere in the world for different customers. Every conceivable kind of robot is being employed, including cartoon characters, an oil can, cigarette packages, kangaroos, a vacuum cleaner, an Easter basket, and a pumpkin. Their client list includes H. J. Heinz, DuPont, Westinghouse, Time, Hoover, 3M, Sperry, and Norelco. One of the company's robots played goalie for two years for the Los Angeles Kings hockey team!

Commencement Speaker

It made news from China to Africa, and Bill Bakaleinikoff's California company, Superior Robotics, gained a lot of exposure back in 1983 when Bakaleinikoff's "Robot Redford" was invited by Maryland's Anne Arundel Community College to be the commencement speaker. A campus furor arose over whether a machine should be given such a distinguished assignment. Reported Susan Spencer, CBS evening news reporter, "Students by and large think the robot is a fine idea but professors are upset." Said one disgruntled

The Quadracon is a popular entertainment robot by ShowAmerica Inc.

professor, "It smacks of being more a college prank than the serious and idealistic event that commencement should be."

In any event, Robot Redford gave a widely acclaimed speech at the college. Actually, as the students were told, it was Bill Bakaleinikoff's speech and his voice, as he operated the robot from off-stage. The topic of the speech was, not surprisingly, the future of high technology. Making light of all the controversy, Robot Reford drew laughter when he quipped, "I understand my coming here today has caused a bit of controversy. I know robots aren't supposed to have emotions, but somebody did not tell my circuit board because the moisture is shorting my batteries."

The Robot Factory

The Robot Factory of Cascade, Colorado, has been designing and manufacturing promotional and entertainment robots for forty-five years. The company currently offers many standard model robots and animated characters. Custom units are also available.

The Robot Factory's first robot, an 8-feet tall giant named Commander Robot, made its debut as an ice skater for the Ice Follies in 1966. By 1968, Commander Robot had been joined by eight more skating robots.

The Robot Factory's first non-skating robot made its initial public appearance in 1971 on the "Merv Griffin Show." It has since been joined by over two hundred "brothers

Robot from The Robot Factory, Cascade, Colorado, greets movie stars.

Animated creatures from The Robot Factory, in Colorado.

"Hairy Three Piece Band," by The Robot Factory of Colorado. Very entertaining!

In 1982, this reward poster was issued by The Robot Factory after someone absconded with "Six T. Robot." Fortunately, a few days later, the robot mysteriously reappeared.

A pretty girl always brings out a robot's personality. Here's the author's daughter, Sharon Kaye.

and sisters" in the United States, Europe, South America, Australia, Africa, Canada, Mexico, and Japan.

The Robot Factory's robots and animated characters are singers, dancers, actors, comedians, radio and television personalities, sales robots, teachers, and therapists. They've appeared on television shows such as "Buck Rogers," "The Jackson Five Show," "Hour Magazine" and "PM Magazine;" done numerous television commercials promoting everything from video equipment to computers, from shoe polish to yeast, and have even endorsed political candidates.

International Robotics

International Robotics is a well-known promotional robot company with its headquarters in New York City and branches in Bangkok and London. The company has taken on projects such as the furnishing of a space age nightclub complete with robots and the manufacture of Disneyland automation characters for Disney's EPCOT Center in Florida.

One of International Robotics' most famous show robots is SICO. SICO has been all over the world; he's met with heads of state, movie stars, and royalty; and he's been on a dozen television shows. The life-size robot's stunning appearance is one reason for his success. His huge, buggy eyes are shocking and draw immediate attention, as does his metallic gold and chrome body. He's had several versions over the years.

Robonosis

Robonosis.com is advertised as a "strolling hypnosis" robot that can walk around anywhere, including colleges, fairgrounds or convention centers. He is intimidating, standing over 6'6" tall, yet kids are drawn to Robonosis. He's a walking photo opportunity.

Kids love SICO from International Robotics.

SICO with the late comedian Danny Kaye and Drew Barrymore as a photogenic little girl. (International Robotics, Inc.)

Ursula, the Female Android

Florida Robotics is a company that makes and rents all kinds of robots, including "Ursula," the female android. Ursula can be dressed up to order. She talks, dances, and more.

Florida Robotics also has "K.I.R.K.," a more traditional (clunky and metallic) robot, "Hard Drive," and many others, each with a distinctive look and a unique personality.

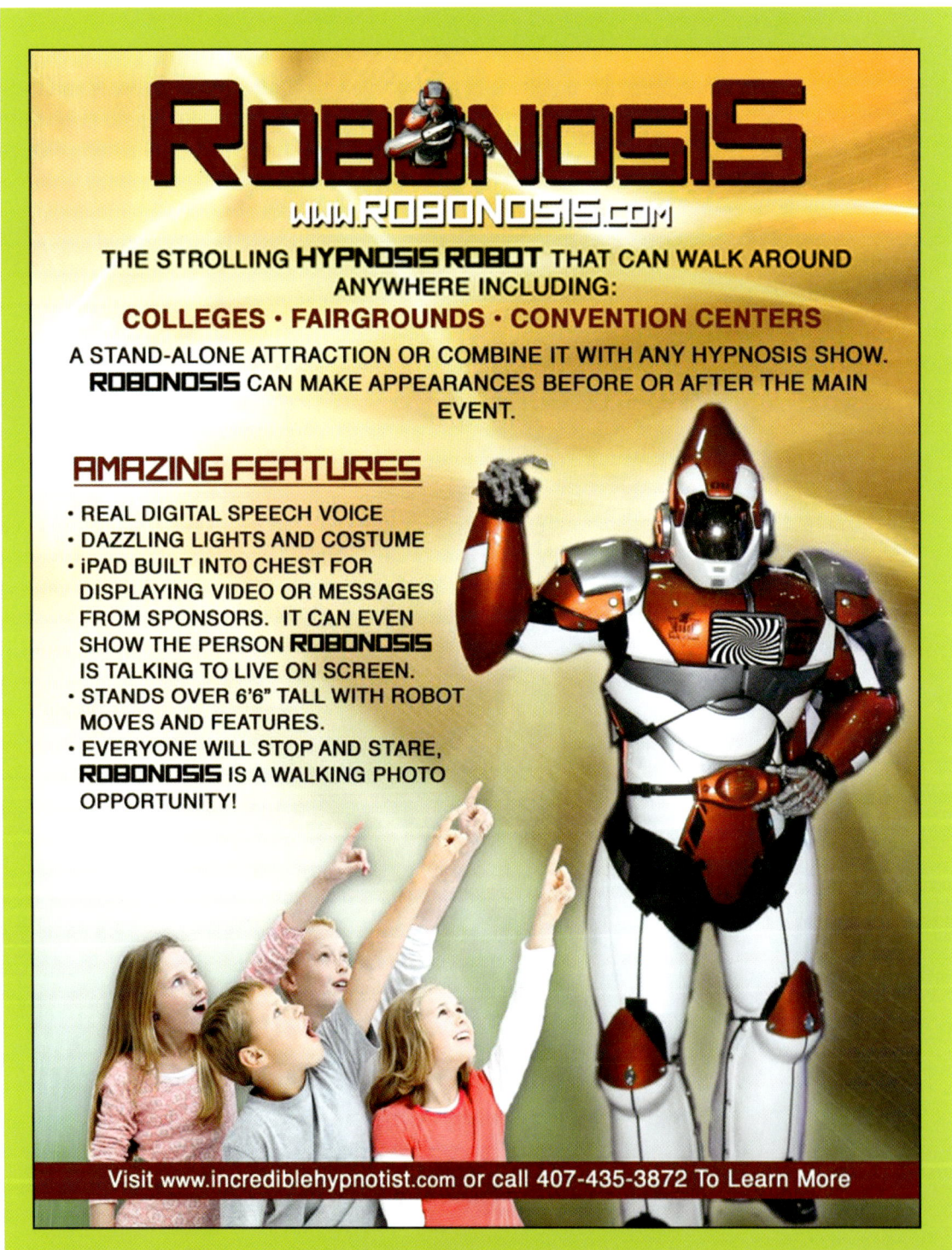

"Ursula," the female android by Florida Robotics, shown here in a movie at Universal Studios. Ursula talks, dances, and more.

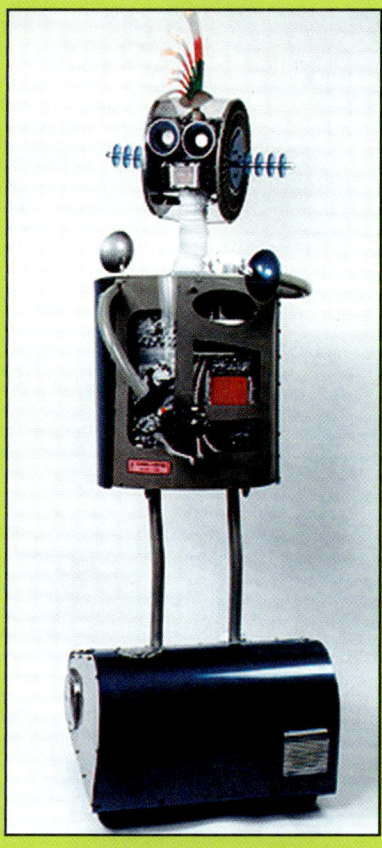

"Hard Drive" (Florida Robotics)

"K.I.R.K." (Florida Robotics)

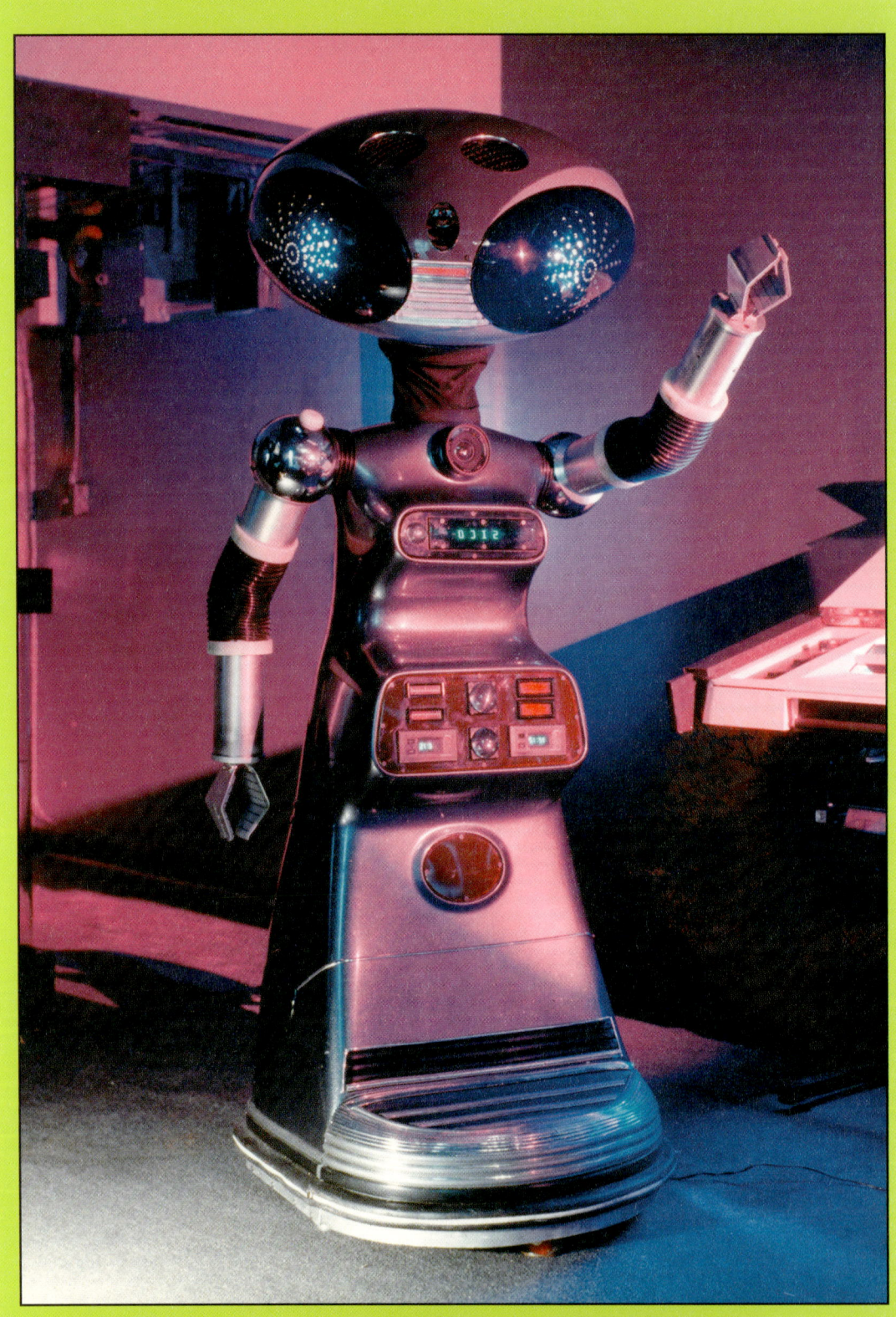
Another version of the popular SICO robot (International Robotics, Inc.)

Robot SICO with former President, Gerald Ford.

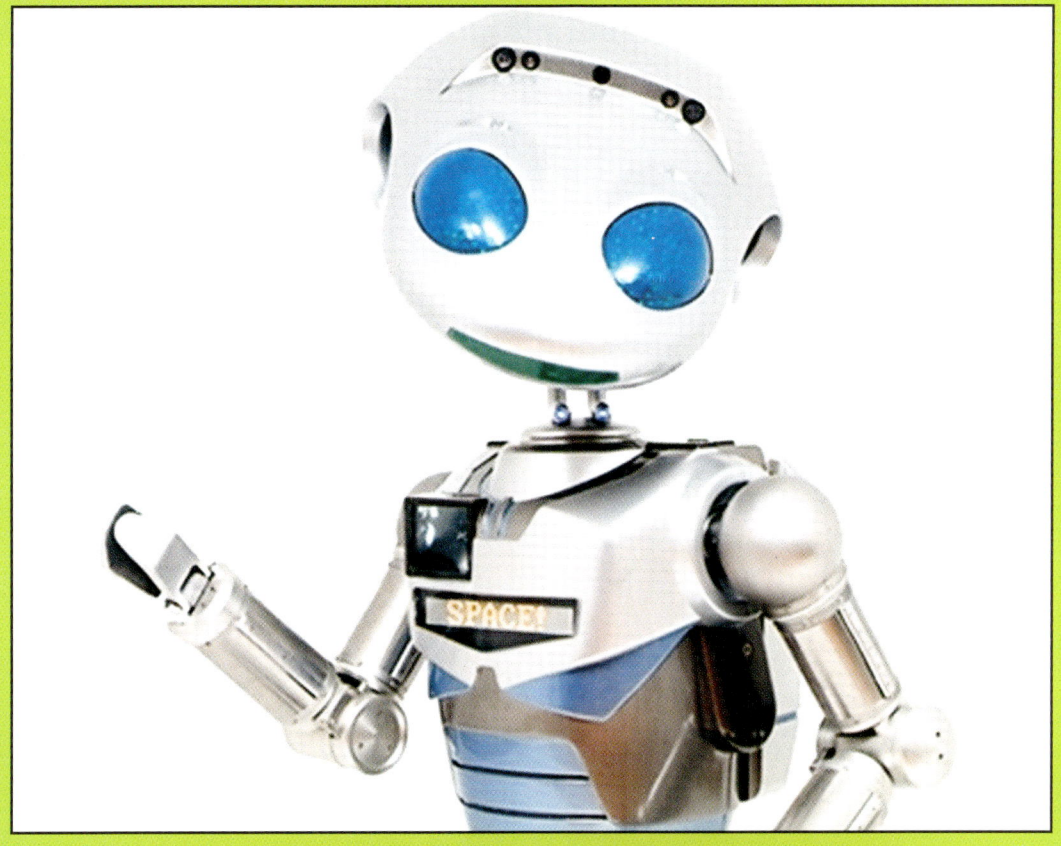
Newer version of SICO. (International Robotics)

International Robotics in New York can custom make you or your firm a robot.

In Japan, Coca-Cola's robot kiosks interact with the customer. The kiosks are so popular they have been made into a miniature "Cokebot" toy for kids.

Chapter 14

Military Robots: Machines to Conquer

Robots are going to war. In an age of high technology, when the computer, robot, and other machines are handling so many tasks and jobs once the sole province of human beings, it makes sense that robots also become electromechanical warriors. The world's top superpower—the United States—is working hard to develop robot systems for warfare and battle.

Billions of dollars have already been spent by the Pentagon on robotics research and development. That figure includes expenditures for robotic devices such as robot underwater craft, cruse missiles—which are, in effect, robot craft—drone aircraft, bomb and mine clearing devices, and robotic soldiers and animals. Military experts point out that "steel is cheaper than lives," and they believe that robots will save thousands of lives in future conflicts.

To gain a perspective of the surge in spending on military robotics, it is instructive to have some of the robots currently on "active duty" with the U.S. armed forces pass in review.

Robots in Khaki and Olive Drab

The U.S. is big on robots. One ongoing project is the development of a tactical robot vehicle. Programmed with terrain map to help it find its way and equipped with computers, artificial vision systems, and other sensory devices, a robot tank will scout enemy territory on its own. The tank is outfitted with firepower so it can lead an assault. "Send it up ahead," says Frank Verderame, assistant director for the Army's robot research program, "and it gives the troops a two or three kilometer cushion."

The Army has several other robot systems in the works or in use. One already in the inventory is a telerobot, an explosive device robot used in the recovery of live fuses and other munitions. Two systems under development are robot ammunition loaders: one loads 45-pound shells in tanks, the other loads ammo for the 105mm howitzer artillery.

The Unmanned Drone (UCAV)

A more exotic robot is the drone, also called the unmanned combat air vehicle (UCAV), designed for use in spy reconnaissance, target acquisition, and guidance of ordnance.

Launched from a ground-launch platform, the UCAV is able to fly and hover over enemy territory; when it spots enemy vehicles or troops, it sends information to distant U.S. Army or other units. The unmanned craft can direct accurate firepower against the ground targets it locates. The Predator UCAV, the Draganflyer X6, and the Shadowhawk are three common types.

The drone is the weapon of choice to find and destroy enemy combatants on the ground. Police departments around the nation are using it, and the CIA and other intelligence units, too.

Security Robots

The U.S. Army is also developing mobile security robots. Some military authorities predict that by 2025, some 25,000 military security guards will be replaced by weapon-toting robots that will guard ammo storage areas, warehouses, aircraft flightlines, classified restricted access facilities, and the perimeters of military installations. Major Clyde Romaster, an U.S. Army spokesman, emphasizes that, "Robots don't fall asleep on guard duty, and they don't need to be fed during a shift. Unlike human soldiers, they are expendable."

Robots in Air Force Blue

Robots are considered a priority in the U.S. Air Force. The air-launched-cruise-missile

Drones (UAVs) come in all shapes and sizes. They are remote-controlled.

An early U.S. Air Force drone.

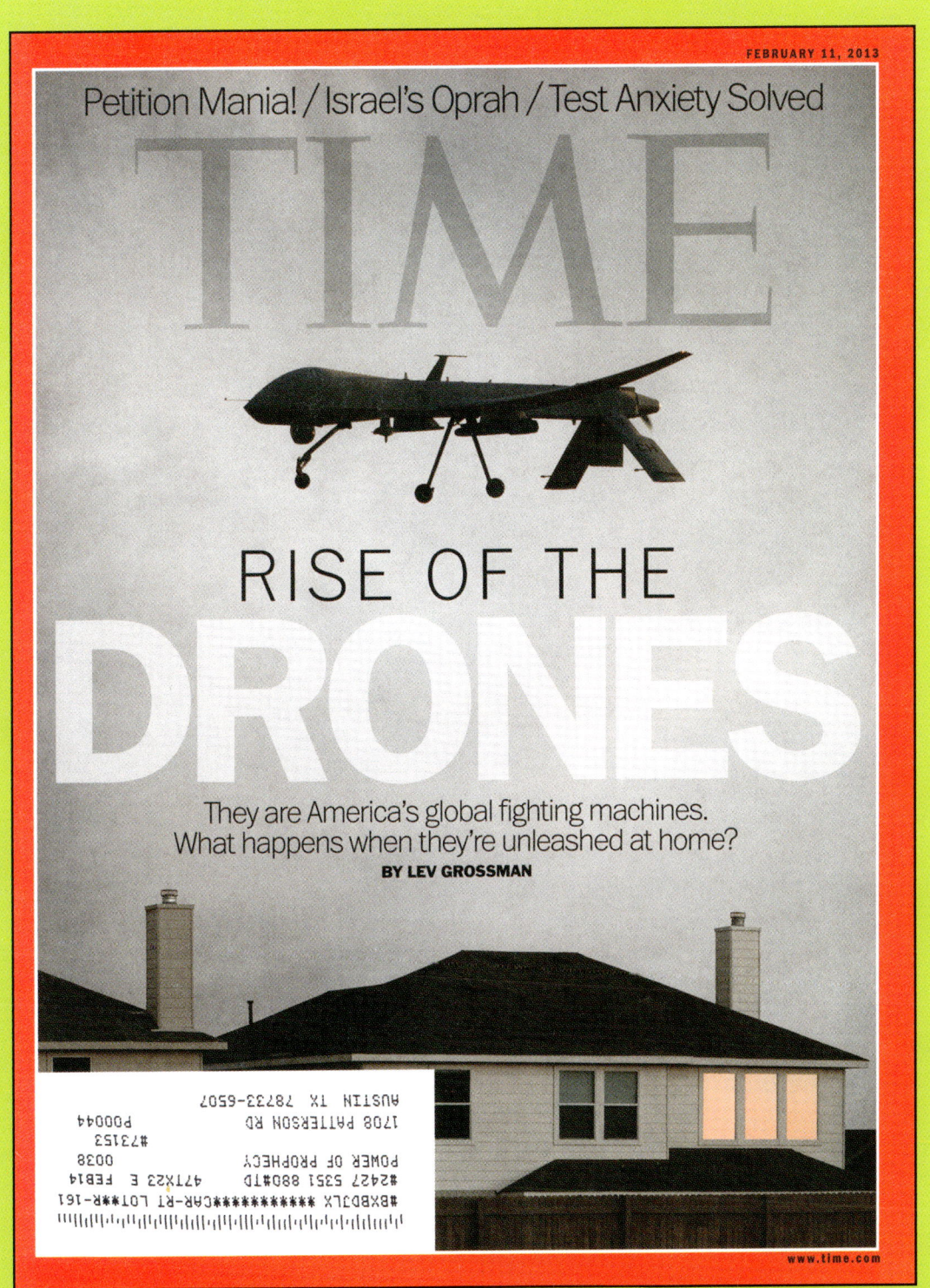

This cover and feature article of *Time* (Feb. 11, 2013) indicates that many are concerned about Big Brother government's robotic drones.

The Draganflyer X6 drone during a test flight in Mesa County, Colorado. Only 36 inches long, it weighs two pounds.

The ShadowHawk drone is used to kill would-be terrorists around the world. Here it is shown with the Montgomery County, Texas SWAT Team.

(ALCM) is a terrain following robot missile preprogrammed to zero in on specific targets. Nuclear-capable and costing only one tenth as much as an intercontinental ballistic missile, the ALCM can change course in mid-flight and even independently choose an alternate target. One sea-launched version, called the Harpoon, can range sixty miles, zipping over the waves before its lethal force impacts against an enemy aircraft carrier or other sea vessel. Another version developed by the U.S. Navy, the Tomahawk, a ground-launched version, is a long range vehicle that can strike a target more than 1,000 miles away.

The U.S. Air Force is also vitally concerned with robotics use by civilian aerospace contractors. Air Force personnel work closely with civilian corporations—Lockheed, Boeing, Bendix, TRW, General Dynamics, and others—to bring robots and other automated processes into aircraft and vehicle assembly lines. Many of the most innovative robotics systems in use today were developed with the use of grant money from the U.S. Air Force and the other armed services.

Scary Autonomous Robots

The military is now building robots that will be autonomous. This is a grave problem in conflict because innocent people may be harmed and killed. Already we've seen "accidents" happen with drones that operate overhead and cannot properly and accurately view the scene below.

Some researchers and elements of the public are violently opposed to autonomous "killer" robots. "If you have an autonomous robot," says Noel Sharkey, Professor of Artificial Intelligence and Robotics at the University of Sheffield (UK), "then it's going to make decisions who to kill, when to kill, and where to kill them…that is really worrying to me."

Nevertheless, the development and use of drones and other autonomous or semi-autonomous robotic weapons continues unabated.

The Future of Military Robots

The Pentagon and the individual armed services see a bright future for military robotics. Robots are simply too useful for military applications to let this technology lay fallow. Robots can go places humans and traditional manned vehicles cannot go. For example, they can enter and work in areas contaminated by nuclear, biological, and chemical weaponry.

The Western European military forces, especially the British, also have plans for robots. The British already have used their "Wheelbarrow" bomb disposal robot in Northern Ireland, where terrorist bombs are a hazard.

Robot Wasps, Flies, and Other "Insects" and Birds

That wasp, bee, dragonfly, hummingbird, or other insect and bird that keeps buzzing and flying around your landscape might just belong to the Armed Forces. The CIA and other

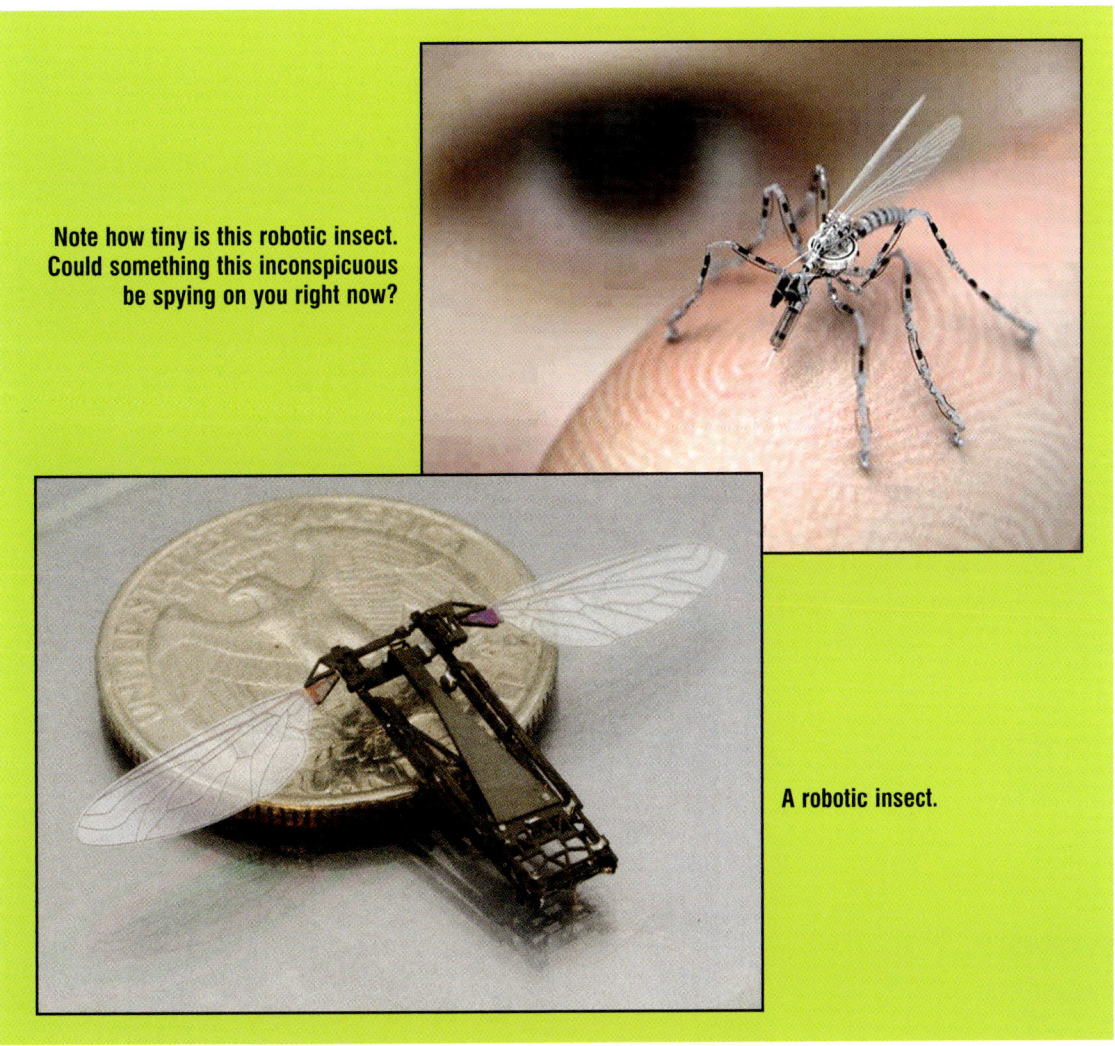

Note how tiny is this robotic insect. Could something this inconspicuous be spying on you right now?

A robotic insect.

intelligence and law enforcement agencies are also experimenting with robotic birds and insects. Some are equipped with a tiny microphone, others with a reconnaissance camera. Teleoperated, they perform spy and surveillance missions. In future conflicts, a great number of these tiny robots, constituting a swarm could be directed at an enemy.

The AlphaDog

The LS3 AlphaDog is a robotic, four-legged military vehicle capable of carrying about 400 pounds of equipment and supplies. Looking like a huge, muscular dog, the "LS" stands for "Legged Square Support System." It's a seeing-eye dog, too, and can travel through rugged terrain for some 20 miles without needing a break.

"The vision for LS3 is to combine the capabilities of a pack mule with the intelligence of a trained animal," explained Army LtCol Joe Hunt in the publication *IEEE Spectrum*.

The huge LS3 Alpha Dog pack-mule, four-legged robot can hike 20 miles.

The Cheetah

DARPA's "Cheetah" robot can run faster than you. It looks like a four-legged *Star Wars* creature, but can it run! The Cheetah gallops up to 28 miles per hour, a land speed record for legged robots.

"The robot's movements are patterned after those of the fast-running animals in nature," says a news release on DARPA's (Defense Advanced Research Project Agency) website. "The robot increases its stride and running speed by flexing and unflexing its back on each step, much as an actual cheetah does.

Titanoboa

The Titanoboa is a Japanese product, a massive, 50-foot, one ton robotic snake. Titanoboa weighs over 2,000 pounds. This robot will support a rider and is supposed to strike terror in an opponent's heart.

Kuratas

Another Japanese robot, Kuratas is a huge 4.5-ton gun that fires 6,000 ball bearing steel

ROBOT ALCHEMY • 197

The "Cheetah" robot is faster than human beings. The four-legged creature gallops up to 18 mph.

The Titanoboa, a huge 50-ton robotic snake.

pellets per minute. Unveiled in 2013, the Kuratas has a diesel-powered engine and its top speed is 6.2 miles per hour. It can be customized for more peaceful functions, too, like firefighting or cleaning.

Crowd and People Control

It's difficult to see how the Cheetah or the LS3 AlphaDog and similar robots can be effectively used for military combat. It seems these fearsome weapons are more likely to be used against demonstrators and protestors, for domestic control purposes.

In the era of Big Brother, we will see robots used against crowds and mobs, to suppress demonstrations. Scientist Paul Marks asks, "how long before we see packs of droids hunting down pesky demonstrators with paralyzing weapons? Or could these packs be lethally armed?"

A Micro Air Vehicle (MAV) sits on a wire awaiting remote instructions. The MAV is a U.S. Air Force product. A drone, it can operate in swarms, hover like a bee, crawl like a spider and even sneak up on unsuspecting victims and execute them with lethal precision. "In the MAV, we downsize weapons and revolutionize warfare," said an Air Force spokesperson.

Chapter 15

Robots in Space

In the summer of 1976, an unmanned Viking spacecraft landed on Mars. Its mission: to search for life. The spacecraft was the National Aeronautics and Space Administration's robot lander craft. Equipped with infrared vision, an on-board analysis laboratory, and a sampling arm, the Viking Lander carried out several important scientific experiments. It dug holes and scooped up Martian soil and analyzed those samples to determine if they harbored—or could harbor—life forms, however minute. The robot discovered that Martian soil did contain the nutrients necessary to nurture life, but no life was discovered.

The 1976 Viking mission was spectacular, but it wasn't the first on which a robot successfully explored the heavens. An earlier Surveyor craft had dug a narrow trench in the moon's soil under the direction of technicians on Earth. For the Russians, it was a robot—not a cosmonaut—that was first to step out—or, rather, roll out—on the lunar surface. On November 17, 1970, the Russian's unmanned Luna 17 craft rolled the

Curiosity Rover spent two years on Mars, looking for signs of life.

Lunakhod I, a space rover. The Lunakhod wheeled vigorously around the rocky lunar terrain sending television images back to Soviet scientists on earth.

Both America's NASA and Russia's space agency have additional plans for robots. For NASA the immediate goal is to further improve the performance of the Space Shuttle's robot manipulator arm and to eventually develop a fleet of unmanned robot servicing craft. The robot manipulator arm has proven a resounding success for the United States. in April 1984 it was extended from the Shuttle's cargo bay to assist in retrieving the crippled Solar Max satellite.

NASA's newest innovation is a giant "megabot" arm called the "Dextre."

NASA's Robots Take Many Forms

In space application, robots may be given a variety of forms, though few look like the popular science fiction conception of a mechanical human. As with industrial robots their appearance usually follows strictly functional lines, varying with the requirement of specific missions. They may take the form of earth-orbiting space telescopes, unmanned planetary spacecraft, or lander craft with manipulators for handling soils and rocks. They may also be wheeled, robot rover vehicles.

However, more recently, NASA has moved into humanoid robots. *Robonaut* is a joint DARPA-NASA project designed to create a humanoid which can function as an equivalent to human beings. Robonaut employs the Segway HT for locomotion, uses telepresence, and uses touch sensors in its hands.

The Space Program also employs industrial robots here on Earth. One strips spacecraft of worn and damaged paint and other exterior coatings. A large arm robot, it blasts the exterior structure with a jet-like spray combination of water and other corrosive material that cleans and restores the spacecraft's exterior to bare metal.

On February 12 and 13, 1977, Viking Lander I dug a series of deep trenches on Mars to provide samples, from as far as 30 centimeters (12 inches) below the surface. This is the actual view from the robot craft.

In 1984, a remote manipulator system arm was used for a series of repair tasks on the Challenger space craft. (Courtesy: NASA)

Dextre, a ten meter wide controlled megabot created by NASA (Courtesy: NASA)

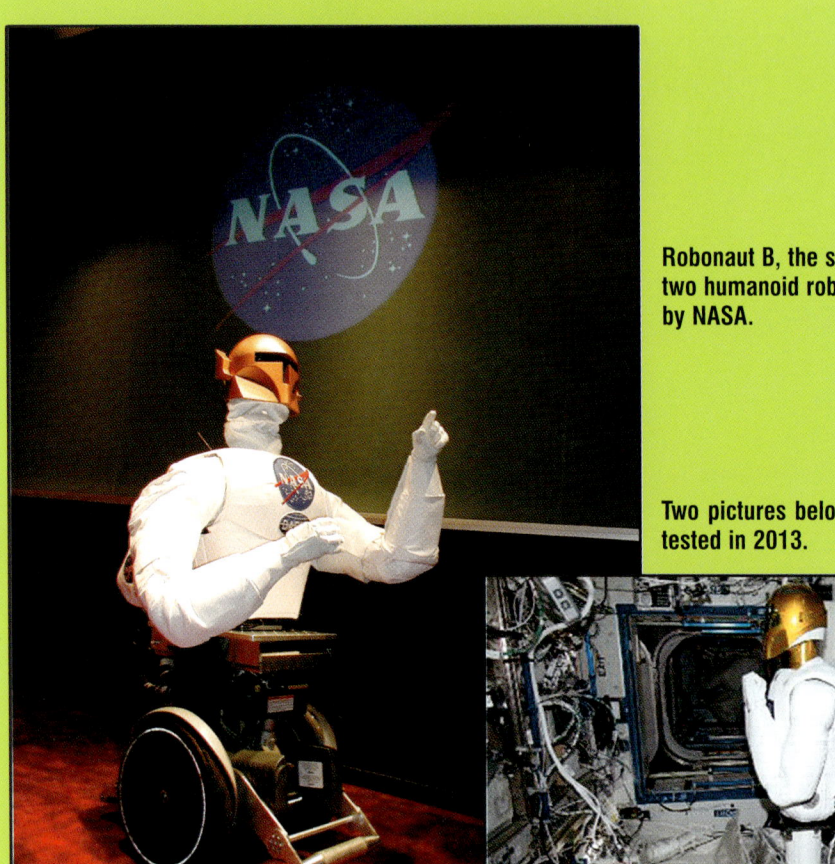

Robonaut B, the second of two humanoid robots used by NASA.

Two pictures below: Robonaut B, being tested in 2013.

Chapter 16

The Robots Will Impact Your Life

Homo sapiens, as he has been pleased to call himself, is, in his present form, played out. The stars in their courses have turned against him, and he has to give place to some other animal better adapted to the fate that closes in. The new animal may be of an entirely alien strain or may arise as a modification of the man species... but it will certainly not be human.

— H.G. Wells

The philosophical issues regarding the robot strike at the very heart of the debate about man's nature—and his future. Expert after expert has been telling us that the robotics era will result in dramatic changes in our work and social lives.

Some would even propose that robots are ultimately the wave of the future: technological evolution's replacement for the human species. Robots, they warn, will evolve into a superior breed that will first throw the biological human race out of jobs, then, finally, entirely off the planet Earth.

This ominous future is not at all what the optimists see as humankind's fate. Instead, the optimists believe the robot will become man's slave, performing our labor and serving our every whim, ushering in a happy, bright epoch. In this new period, *homo sapiens* will be freed by robots from the drudgery of performing physical labor and making drab calculations. Thus, humans will have time to exercise their native creativity and imagination. The robot, say the optimists, will prove to be humankind's salvation, not his downfall.

As momentous as this issue of man's—and robot's—ultimate future is, let's shift our focus to the more immediate future; say, the next ten to twenty years. It's much easier to examine this medium-range future than to engage in ultimate speculation and prediction.

What can we expect from robots in the coming years? Among the important questions that deserve our attention are these three:

A long time ago, in 1906, before electric appliances, an artist came up with the idea of an unusual home service robot that could be operated by steam. Contrast this robotic servant with the snazzy model at right.

- How soon will we have robots in the home? What will they be able to do for us?

- How will robots affect our economy? Will industrial robots throw millions of manufacturing and service workers out of jobs?

- What is the shape of robots to come? What kind of robots will we be coexisting with? What will they look like?

In exploring the answers, a great many experts have been polled and much of the available literature has been examined. History has shown that expert prediction is not prophecy, but the informed opinions and vision of Joseph Engelberger, Marvin Minksy,

Isaac Asimov, and others quoted here deserve our consideration. They've been amazingly on-target in so many of their past predictions that their forecasts of the exciting future merits our rapt attention.

Enter the Robot Home Companion

The robots we can now take home are of limited utility, but those who've reason to know confidently predict that more useful home robots are just around the corner. If the experts are correct, in our generation we should see humanoid robots that perform most of our onerous household chores, serve as pets, carry on animated, interesting conversation, and educate the kids.

Doug Bonham, Marketing Director of the Electronics Division of Heathkit says, "In twenty or twenty-five years, most of us will have robots handling chores in our home. Personal robots will obediently take over menial chores and become reliable home appliances much like today's washing machine or vacuum cleaner."

Robotics author Nelson B. Winkless, III, has a vision, too: "Home robots will eventually be able to talk to you. They will be outfitted with medical diagnostic sensors and keep track of your physical condition. Your robot, Eunice, will sidle up to you and say, 'You know, your blood pressure is a little high. Maybe you better call Dr. Fishback.' When Dr. Fishback arrives, he talks to you *and* Eunice the robot. He asks for statistics—which days your blood pressure was high and so forth. Eunice dutifully recites your record with digital precision."

"Doberman robots could replace a real canine as a home or business watchdog," says Lawrence D. Gasman of International Resource Development, "and robots will assume household roles as their intelligence increases."

Here's what other experts have to say about the future of home and personal robots:

This robot hand isn't pretty, but it gets the job done. Still, would you want to shake hands with this guy? (Pyramid Films)

The home of the future will have a robot for every room—a kitchen-based robot to plug into increasingly sophisticated appliances, an entertainment robot to activate television screens, and a garage-based robot to mow the lawn.

— George Cretectos
Owner, Robotland

It will take some 25 years to develop fully integrated domestic servants. Meanwhile, many people will be startled in the next few years seeing $5,000 robots vacuuming airport lobbies or mowing the neighbor's lawn.

— James L. Crowley,
Researcher
Carnegie-Mellon University

I believe we have the technical know-how to produce [domestic] robots by the beginning of the next decade... Ideally what you do is build a home that is compatible for a robot... The robot would handle all cleaning and vacuuming. It could even wash windows... be used as a baby-sitter and as a warning system to alert the family in case of fire or other emergency.

— Joseph F. Engelberger
Founder, Unimation

Robots will be programmed to fulfill such household duties as tutor, chess partner, watchdog, psychoanalyst, accountant, lover...

— Neil Frude
Robotics Author

In 5 to 10 years we will have full-featured robots that will allow us to "love the one you're with." Forty years later, we'll have fembots that we can fall in love and have a relationship with.

— David Levy
Author of *Love and Sex With Robots*

Robots, Jobs, and the Economy

Robotics and automation are rapidly transforming the world's economies. Harry Mackie of GCA Corporation believes that the factory of the future, integrating advanced forms of automation and robotics, is the *only* factory *with* a future. However, not only large factories of huge corporations will undergo automation and robotization. That's just he first step. Small workshops, offices, retail stores, and most other segments of the economy will see robots march in, too.

Service industries will not be exempt, either. The Research Institute of America recently reported that, "mobile robots that cut grass, clean floors, load dishwashers, work as prison guards, report and put out fires, paint walls, and blow snow from clogged driveways are all under development."

In Japan, the Ministry of Manufacturing Trade and Industry has begun a seven-year program to develop robots that can carry out a variety of non-manufacturing tasks.

Lego's Robots: Here are just two of the seventeen programmable robots that adults and kids get when they purchase the Lego Mindstorms EV3. Lego is the same company that makes the famous construction kits. Since 1998, Lego has produced robot teaching tools. The latest (2013) is Lego's *Mindstorms EV3* platform. This third generation Lego robotics platform provides hobbyists and enthusiasts with the capability to construct seventeen programmable robots. These are actual robots that feature sensors, motors, and movement. Each robot is different and guarantees the owner many hours of adventurous learning—and great fun.

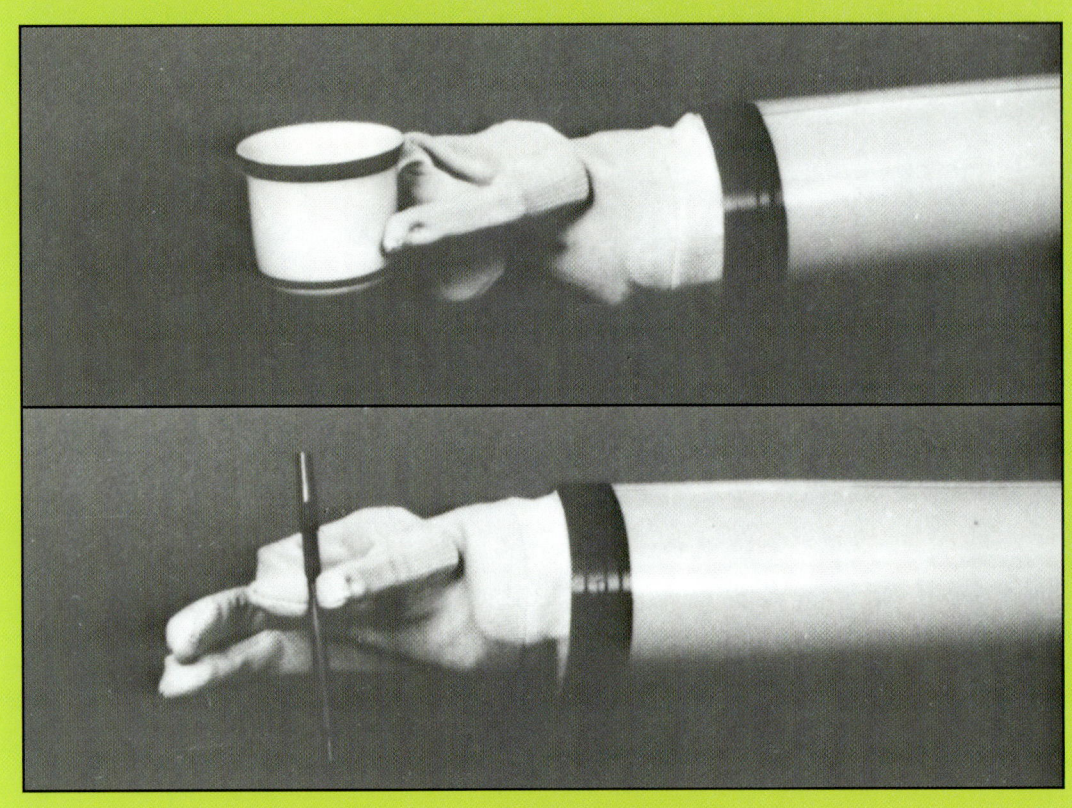

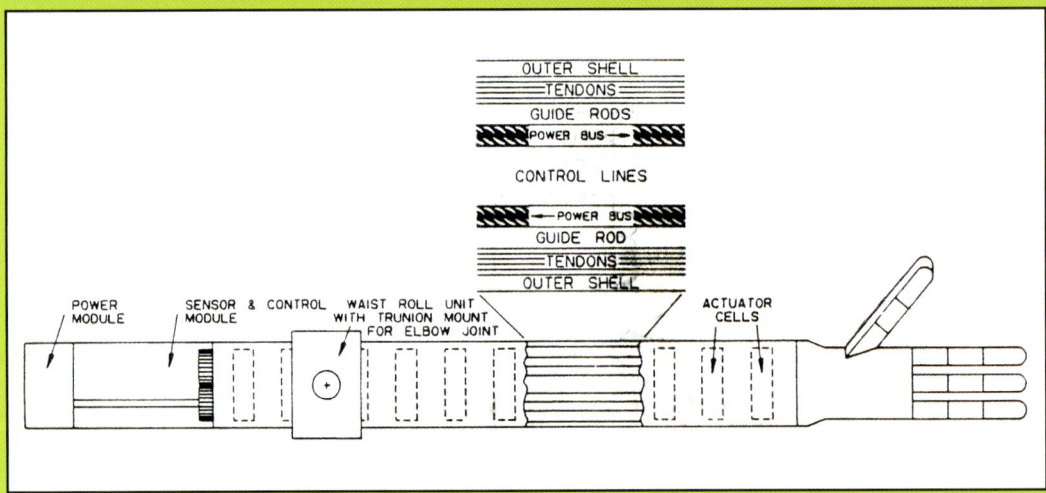

This is a picture and a detailed drawing of the robot forearm and end-effector (hand) by inventor David Peterson of Victory Enterprises Technology.

These tasks include inspecting eggs, spreading fertilizer, sewing garments, guiding the blind, sweeping streets, and acting as nurses in hospitals.

Two industries that soon will experience dramatic changes because of robotics are construction and agriculture. Construction robots are close to technological reality, and they will soon be economically justifiable. In Japan in a test project, an entire 1,600 square foot home was built by robots alone. The robots welded, tacked, cut, glued, and painted. The three-bedroom, two-bath prefabricated home was built in an astounding forty minutes on an assembly line and was assembled on site in only four hours. Your own house may have taken months to construct, but the Japanese robots completed an entire home in a total of four hours and forty minutes.

Down on the Farm: Old McDonald Robots

If you can't keep the kids on the farm because city lights are so captivating and good farm help is hard to find, there's a solution: robots. Robots are increasingly showing their stuff in the agricultural industry.

In Florida, the Martin-Marietta Corporation is developing a robot system for citrus picking. The orange, grapefruit, lemon and lime picker machine has a computer brain and a photocell array that allows it to figure out which fruit is green and which is ripe for the plucking. The ripe fruit is grabbed by the robot's arm and hand and neatly deposited in a harvesting tray.

All over the world, engineers and other experts are working on farmer robots. In Montpelier, France, a team led by agricultural engineer A. G. d'Esnon has also built a prototype apple- and fruit-picking robot that can select and pick only the ripe fruit. French engineers also are testing robots that prune grapevines. To give this a humorous dimension, can you imagine wine grapes trod not by pretty young damsels, but by robot feet? Even the most dedicated roboticist might protest. Romance, after all, must count for something.

Ranchers may also benefit from robots. In Australia, there's a robot that does a remarkably capable job at shearing the wool from sheep. And in Texas, several owners of cattle ranches are studying the use of robots to brand cattle and keep a watch out for rustlers.

Experts see a bright future for robot farmers. At the International Computers in Engineering Conference and Exhibit in Las Vegas, agri-authorities predicted that robots will soon be performing such down-on-the-farm tasks as milking cows, cultivating crops, fumigating to get rid of insects and pests, and furrowing and tilling land. Several different types of robots will be used by farmers, including stationery and roaming models, robots that can see, and robots with human-like hands and a keen sense of touch.

Robots and Unemployment

We've already heard some of the forecasts that robots will eventually cause mass unemployment. Pro-technology people deny this, pointing out that, historically, technology has produced more jobs. They admit that robotics will result initially in job loss for many workers, but they believe that on balance, new, different jobs will be created as the economy benefits from and assimilates robot utilization. However, robotics seems to be of a different fabric because robots can be created to build more robots.

Jobs With Robots

While many workers are being replaced by robots, others are finding that, temporarily at least, the Robotics Age is bringing with it fantastic career and job opportunities. According to Walter Weisel, former president of the Robotic Industries Association, ironically, "The one holdback to robotization is a skilled human workforce."

Weisel says that workers are in great demand to design, manufacture, sell, install, maintain, and service the new scientific machines. Adds Professor J. P. Lisack, director of

Odetics produced this powerful walking humanoid.

Purdue University's Office of Manpower Studies, "There are excellent career opportunities for many people in robotics to keep our manufacturing and production facilities operating smoothly... Qualified workers—doctors—are needed to engineer and service the thousands of new robots."

What are the most promising jobs in robotics? Although people in a variety of job fields are needed, *robot technician* is expected to be the number one growth specialty over the next decade. Other specialties in demand include *robotics engineer, manufacturing engineer, industrial engineer, electrical and electronics engineer, mechanical engineer, electro-optical engineer, biomedical engineer, CAD/CAM specialist, laser technician, engineering technician, computer programmer,* and *robot programmer*.

The number of vocational-technical schools, colleges, and universities offering degrees or programs in robotics is growing rapidly.

The robotics industry itself may produce thousands of new jobs. Bruce Nussbaum, editor at *Business Week* magazine and author of *The World After Oil* (McGraw-Hill,) predicts that "The robot industry will soon become one of the largest, most profitable and most important industries in the world." Robotics, Nussbaum says, "will dominate our economies, much as autos, chemicals, and steel dominated the 1960s and 1970s."

Eventually, however, the growth of robotics into a dominant industry may not, after all, compensate for jobs lost. Self-replicating robots—robots that build other robots—and artificial intelligence software might preclude the employment of a large human workforce in robot factories. Past technologies, including the advent of the Industrial Age, cannot be reasonably compared to the radical restructuring of world economies that might result from automation and robots.

A Congressional Office of Technology Assessment report forecast high joblessness in the northeast and midwest United States in future years as industries become more automated. Declines were forecast for craft workers, laborers, and clerical personnel.

There are already indications that automation is causing unemployment. In Japan, unemployment over the past decade has doubled, and many analysts attribute the rise in joblessness to the growth in robotics during the same period.

Robotics may also severely damage the economies of the emerging countries. Robots can do jobs cheaper and with higher quality than humans—even humans paid the dismal wage rates that prevail in many nations in South and Central America, Asia, and Africa. Increasingly, the industrialized nations will bring assembly plants back home, where robots can be used to best advantage. No longer will it be financially advantageous to use cheap foreign labor or to seek tax shelters overseas.

In summary, whether or not robotics will result in long-term mass unemployment is an open question. But it is a certainty that in the short run, robotics and automation will cause economic dislocations, force the retraining of millions of human workers, and leave those who can't be retrained permanently unemployable. Experts say that industry and government should be planning *now* for the economic turmoil that lies ahead so the problems can be handled as smoothly as possible. Douglas Oleson, chief operating officer of Battelle Memorial Institute, has said, "It should be our goal to make certain the change is for the better—that technology creates a decent standard of living for people everywhere."

In 1966, General Electric was commissioned by the U.S. Army to design this walking humanoid. Impractical, it gulped down 50 gallons of oil per minute.

Ivan Sutherland of Carnegie-Mellon University produced this Cybernetic Walking Machine.

The Shape of Robots to Come

Robots in the late 2020s and beyond will be physically and mentally very different from those of today. In factories, intelligent robots will be built to accommodate function. Increasingly, such robots will be viewed more as machines than as robots and they won't resemble humans. Agricultural robots will look like mobile machines with arms and hands. Tomorrow's robot airplanes, automobiles, jeeps, and tanks will appear just as do today's. The only difference will be that they will be unmanned, and drone aircraft will be everywhere.

Many robots for the home and at least some robots in service industries will resemble the human form. English robotics author Neil Frude suggests, "Computers and robots will be built in such a way that we will be forced to admit they have presence, personality, and charisma." Such robots may have rubberized or soft, plastic-like limbs that lend a human appearance. Indeed, the more advanced home robots may be made to order, with customers specifying they want a robot that looks like a movie star or entertainer.

Robert Malone, author of *The Robot Book* and a respected authority on the sociology of robotics, believes that future home robots will not be just plain-looking, bare essentials machines. Instead, Malone suggests, "They will have to have some esthetic appeal. They'll have to blend in with the environment. Maybe they can be decorated in a French Empire style or have brass fittings and inlaid wood."

They will, of course, until the era of the intelligent robot, when robots become companions and sex partners for humans.

Robots to Get Smarter

Perhaps the greatest change for robots will be their growing brainpower. Although the late Isaac Asimov likened future robots to common tools, emphasizing that "the human brain is made for intuition, insight, fantasy, and imagination," brainy robots may one day possess these same abilities. Already, a computer has written free-flowing poetry. One computer, named Racter, even wrote an entire book of essays and assorted ramblings. The book, *The Policeman's Beard Is Half-Constructed*, was published in 1984 by Warner Books.

Artificial intelligence software and new, highly advanced computer chips—some made of biological material—will expand the minds of robots. Marvin Minsky, the famed robotics and artificial intelligence expert from Boston, maintains that computer circuits will one day be able to produce not only intelligent behavior but also creativity and emotions.

If Minsky is right—and many experts believe his prediction will eventually be borne out—robots may put more people out of work than we now envision. Along with laborers and craftspersons, high-level workers in fields requiring mental activity may also be replaced by robots. One can imagine the implications.

It is logical to speculate that within twenty to thirty-five years robots superior to humankind both in physical and mental capability will be common. Walter Hankins III and Nancy Orlando, of the automation technology branch of NASA's Langley Research Center, told the conferees at the International Computers in Engineering Conference and Exhibit that robots and humans will ultimately work side by side. "In general," the two space robotics researchers said, "robots will work for men, but there may be exceptions in which robots are higher in the hierarchy than some humans."

Eugene Jarvis, a computer and electronics professional best known for designing the special effects for video games like "Robotron" and "2084," theorizes that in a

The author of *Robot Alchemy* and his wife, Wanda, visited by a friendly robot at their home in Austin, Texas.

Predator, a drone that can be armed with deadly missiles, is being used by the U.S.A. to kill enemies far below.

This robotic drone has excellent hovering capability.

hundred years robots may decide humans are inefficient and should be destroyed. Until that late date, though, Jarvis says robots will be extremely useful to have around. "Robots will do all the nasty jobs," Jarvis claims, "and let humans sit and think."

Some experts say that the future will see *teleoperated robots*. Hook up your human mind to the computer brain of a remote teleoperator robot and you can feel, see, and do things through the limbs and eyes of the robot. The robot can tour Rome or Paris or attend a meeting for you while you view and feel everything from the comfort of your living room at home.

Given the rapid pace of advances in brain-wave technology, this is not a far-out notion. An AT&T special report on future technology stated, "We can imagine a technology of 'telepresence' that uses robots equipped with various sensors to transmit perceptions back to the user."

Robot magazine is a current publication highly recommended for its great articles on robotic developments. It is read by professionals and enthusiasts alike.

Is There a Robot in Your Future?

As you can see, robots and robotics are growing in importance. Within a few decades, the world population of robots may zoom a hundred-, ten thousand-, or even a million-fold. Many of these robots will look like humans; they'll talk, hear, walk, see, and touch as we do. Eventually, they will also think and feel as do humans. The human-robot distinction will fade away.

Is there a robot in your future? No doubt, there is. Unless you choose to exist as a hermit. Robots will soon invade almost every waking—and sleeping—hour of your existence.

How will this "invasion of the robots" influence your life? The answer is really up to you… and to all of us. As the great thinker Havelock Ellis once wrote:

> The greatest task before civilization at present is to make machines what they ought to be, the slaves instead of the masters of men.

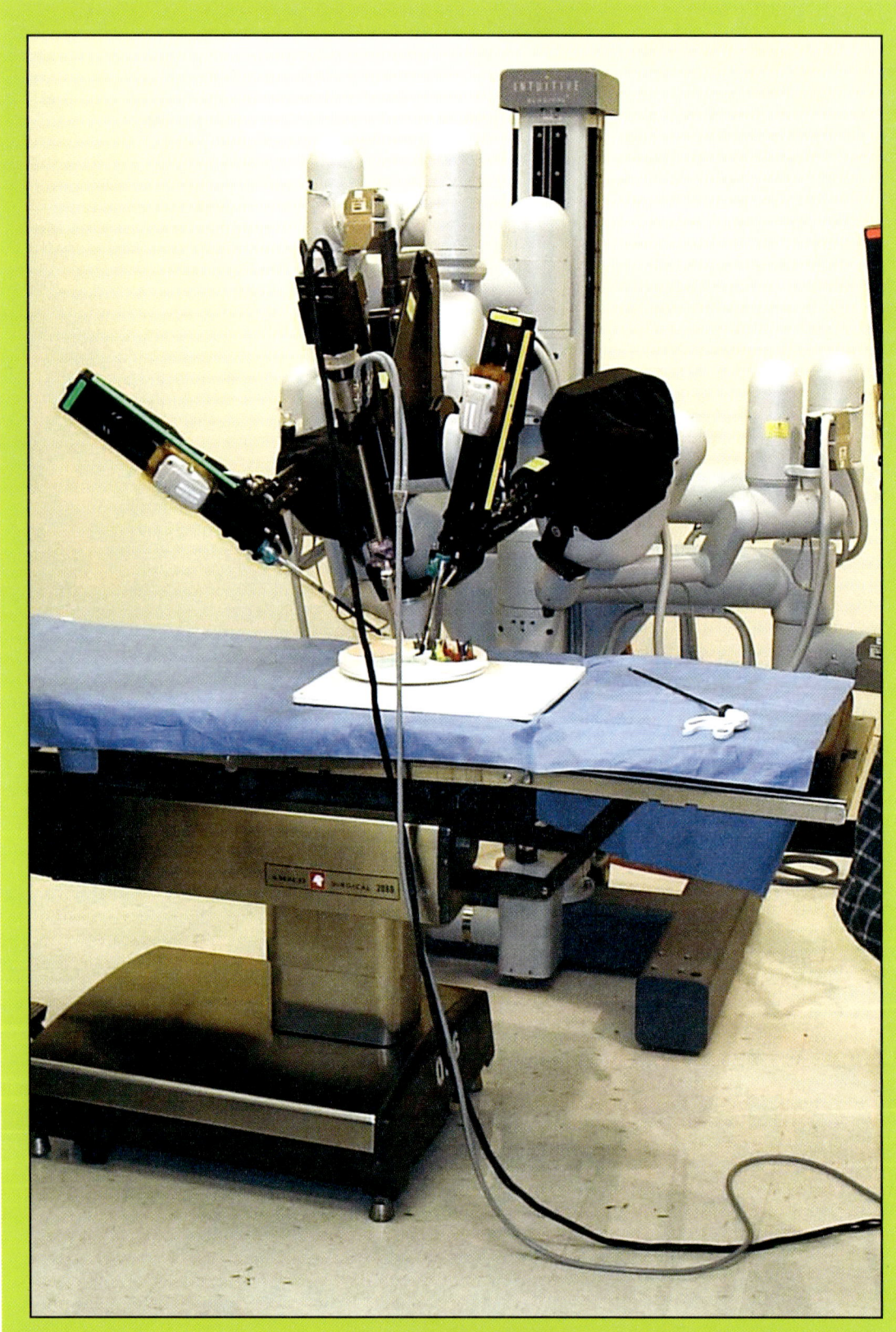
Laproscopic surgery robot.

Appendix I

Understanding Robots

The robots are here. A new species is among us, and the world will never be the same. Hundreds of thousands of robots now inhabit the Earth, and they are doing some remarkable things. Some can talk, walk, calculate math, and even do a little dancing. One is endowed with "emotions." Play a game of tic-tac-toe with this guy and, if he loses, he lets out a loud shriek. But if he wins, he spins around and shouts with glee. Other robots work on police SWAT teams, as waiters and waitresses in restaurants, and as TV and movie entertainers. Robots are replacing human workers on factory assembly lines and helping to retrieve satellites in space. Who knows where it will all end?

Until 1921, no one had used the word "robot." That year, Czechoslovakian playwright Karel Capek wrote a drama, *R.U.R. (Rossum's Universal Robots)*, in which mechanical creations killed all the humans and took over our planet. Capek used the Czech word *robota*, meaning "serf" or "slave-like work," to describe the new, mechanical life forms.

The coined word stuck, and so today the word "robot" brings instant recognition. That recognition can take many forms, though. To a few, the word robot conjures up frightening images of evil, metal, people-like creatures bent on mayhem and destruction. However, many of today's generation are likely to envision as rather pleasing the antics of R2D2 and C-3PO, the friendly droids of *Star Wars* fame.

Some people view robots as neither inherently good nor evil but merely as functional machines. Say the word "robot" and the picture that flashes in their minds is that of a huge, powerful industrial robot working its arm off on an auto assembly line.

All of these images of robots have merit, but they are also incomplete. The technological universe encompasses robots of all sizes and shapes. It includes science fiction and movie robots that are evil as well as those that are good; real-life personal robots that are nearly as warm and friendly as R2D2 and C-3PO; and real-life worker robots with impressive skills but little or no personality. There are tall robots and squat robots, robots with no arms and robots with six, dumb robots and intelligent.

Of Androids, Cyborgs, and Other Robots

People have different ideas of what robots are because, in fact, there are many different kinds of robots. Robot terminology may add to the confusion. For example, do you know the difference between a *robot* and a *cyborg?* How about a *drone* and an *android?* An *automaton* versus a *humanoid?* As you can see, the word "robot" is a generic term used to describe humanlike machines that, technically speaking, go by different labels. However, to properly understand robots, it is important to more precisely define each of these labels. Let's start with the word "robot" itself.

Robot

In simplest terms, we can define a robot as a machine that is humanlike. This implies that the robot can do something automatically that humans also can be expected to do. It may be able to move around, pick up and transport objects, or speak. Note that the robot does not have to look exactly like or resemble a human. This is why industrial arms—machines that are shaped like construction cranes—can properly be called robots. They perform tasks that otherwise might be the province of human workers. Nor are all personal robots human-like in appearance. A few resemble trashcans, others can accurately be described as mobile boxes. Still others could be mistaken for golf carts.

To be considered a robot, a machine does not necessarily have to possess a built-in computer brain. For example, a robot may be remote controlled via a wireless electronic control unit or a remote computer monitor. Some robots are called *teleoperator* devices because a human operator electronically directs their movement from a distant source. Teleoperator devices include mobile robots that dispose of hazardous wastes and also submersible robots that roam the ocean floor searching for ancient shipwrecks. In both cases, a human operator controls the robot's movement while remotely viewing the actions of the robot.

A number of *roboticists* (experts or authorities on *robotics*, the science of robots) believe that the term "robot" should be more restrictive. They contend that a true working-robot must be intelligent, independent, and autonomous, able to sense and interact with its environment, and act according to the feedback received. In essence, this means that a robot must be a machine controlled by a computer and endowed with sensors allowing it to evaluate what is going on around it. Thus, the Robotics Industries Association, a trade group, has adopted this definition:

> "A robot is a reprogrammable, multifunctional manipulator designed to move materials, parts, tools, or specialized devices through variable programmed motions for the performance of a variety of tasks."

This definition may be appropriate for the hard-core roboticist but the global population cannot be expected to applaud the purists. Ask a kid what a robot is and he/she is likely to simply respond that it is a "thing that looks or acts [or both] like a human."

Admittedly, this is a vague definition. For example, is a washing machine a robot? It does, after all, act like a human by washing clothes. Or how about an automatic milking machine or an automated bowling pin setter? These machines perform independently, make decisions, and accomplish work that a human can do. But most people would probably chuckle at the mere suggestion that a washing machine or milking device is a robot. The average kid or adult would probably exclaim, "A robot is a robot, and those things *are not* robots."

Androids

An *android* is a special form of robot. Whereas other robots may resemble beetles, bottles, or even walking locomotives, an android is anthropomorphic; that is, humanlike in appearance. In the 1930s, the first robot builders conceived of robots as artificial people and they tried to construct androids that were as close as possible to human form. Thus the android usually has a head, chest and waist, arms and legs.

Some androids built by robot hobbyists are so realistic they are frightening. It seems that people are more willing to accept a shiny, rotund, machine-looking android with blinking lights and jerky motions than they are an android closely patterned after the human anatomy. Eyeballs, ears, and teeth on an android often prove quite startling.

A *humanoid* is an intelligent android that closely parallels the human form. Robotics scientists have not yet been able to create an intelligent humanoid, and it is unusual to find a crude machine with two legs, two hands, and fingers. More common is the *functional android*, which resembles a human but whose physiology may contrast sharply. For example, a functional android may use wheels instead of legs for mobility and may have an oversized head or body. It may also have claw-like appendages or grippers instead of fingered hands, and its eyes may either be slits or not exist at all.

A special type of non-mobile android is used for certain important tasks. For example, the University of Southern California School of Medicine uses *Sim One*, a computer-controlled mannequin patient as a simulator to teach future doctors some of the skills they must learn. Sim One, the simulated patient, looks amazingly like a living person; it "breathes" and has a normal pulse and heartbeat. When receiving medication or being operated on in surgery, he reacts as would a human.

Another android, called *SID (Side Impact Dummy)*, is used by the National Highway Traffic Safely Administration in simulated auto crash tests to evaluate the impact on real human bodies.

Functional androids may also be given the title of *mechanoid*, indicating that the creature, while roughly approximating the shape of a human, is more akin in looks and function to a mechanized machine. Odetics, a well-known engineering company in California, built a rugged, giant robot the firm calls a *functionoid* because of its ability to perform a variety of tasks and functions—some of which are either too hazardous or require too much lift-power and brawn to be performed by humans. With its long skinny legs and round capsule body, the Odetics functionoid looks more like an ungainly spider than it does a human.

A form of "smart" android that possesses a high level of intelligence and can communicate is called a *droid*. The droid is a robot totally subservient to humankind, a willing slave to a biological master. Droids are loyal helpmates who perform specific tasks for which they are specially built. Often, the droid is a diminutive fella whose physical dimensions are inferior to those of humans. Thus, he is the least threatening of the androids.

George Lucas, in his *Star Wars* saga, gave us the droids R2D2 and C-3PO. And in the 1971 movie, *Silent Running*, actor Bruce Dern lived in space and was befriended by three droid companions, Huey, Dewey, and Louie, who conscientiously watered and tended plants in a greenhouse. The droids also played a mean game of poker.

Automatons (also called *automata*) are not generally considered true robots. An automaton is a representation of a living creature, but it is limited in ability and can perform only a few tasks. An automaton cannot choose a proper course of action, nor can it be controlled by a remote human operator. Examples of automatons are birds in cuckoo clocks, fish in a sculpture that spout water, and toy soldiers that can be wound up and made to walk.

Automatons can be made to look like people and can be quite convincing in appearance. At Disneyland, in Anaheim, California, an automaton that looks like Abraham Lincoln thrills spectators by rising from a chair and reciting the Gettysburg Address, while across the continent, at Disneyworld's EPCOT Center in Orlando, Florida, is "Mark Twain," another automaton. Disney officials call their creations "audioanimatronic devices" and insist they are not robots.

Cyborg

Cyborg (CYBernetic ORGanism) is a mixture of human and robot. The classical cyborg is a human being modified by the addition of artificial limbs or organs. The once popular TV shows "The Bionic Woman," starring Lindsay Wagner, and "Six Million Dollar Man" featured a woman and man whose artificial limbs and organs were so superior to human biological parts that the resulting cyborgs possessed superhuman powers and were able to perform incredible feats of strength and endurance.

Today, the term cyborg is giving way to the phrase *bionic human*. The rise of the science of biomedical engineering, or bionics, and the implantation of electronic and computerized devices—such as artificial hearts and pacemakers—in humans has given the bionic human a huge boost of publicity.

In addition to the cyborg and the bionic human, we have several variations such as the *cyberman* (a robot with a human brain), *cybot* (a completely mechanical robot that has the same mental capability as a true cyborg), and the *cybert* (the perfect cybot: a masterful blend of electromechanical abilities and high level intelligence). Cybermen, cybots, and cyberts are now only science fictional characters. The cyborg, however, is on its way to reality, because of the rapidly developing technology of bionics.

How Many Robots Are There?

No one really knows how many robots there are in the world. But one thing is for sure: we are in the midst of a population explosion. Humankind is entering the Robotics Age with all dispatch, and robots are being spawned at a record pace. In one factory near Mt. Fuji in Japan, robots work twenty-four hours a day, seven days a week, to produce other robots. They do their work with minimum human supervision—and during the night hours, only a single human robot tender is around to watch over their tireless labors.

Are Robots Alive?

People actually are beginning to seriously ask, "Are robots alive?" That's clear proof of just how far robotics technology has come. Many experts in robotics, computers, and artificial intelligence have already concluded that some of today's more sophisticated robots may indeed be alive.

James S. Albus, director of robotics and automation at the National Bureau of Standards, admits, "There is a sense in which it can be argued that robots are an evolving life-form." M.I.T.'s Joseph Weizenbaum agrees. Says Weizenbaum, "When I say, therefore that I am willing to regard such a robot as an 'organism,' I declare my willingness to consider it a kind of animal."

Geoff Simons, Director of the Computer Centre in Manchester, England, has said that, "The most reasonable definitions of life admit the possibility that certain types of artificial systems may be alive." He includes robots in his definition of a living species.

People's reaction when they encounter a walking, talking robot provides some evidence that most of us aren't quite sure if robots are alive or not. At one technology fair in San Francisco, a remote-controlled show robot rolled up and down the aisle confronting people and asking them questions like, "Why are you shaped so funny?" and making bold statements such as, "I'm smarter than you, I have a computer for a brain." The spectators were jolted. Said a little boy to his dad, "It's alive, isn't it?" One woman went up to the robot, whispered something inaudible in its ear, and kissed it. "Wow!" exclaimed the android, "It's a date!"

RB Robot Corporation, the manufacturer of the intelligent RB5X personal robot, offers its customers auto bumper stickers that assert "Robots are people, too." Says Sharon Smith, marketing director of the company, "Many people believe that our robots are living creatures. Some kids prefer RB5X to cats and dogs. "

Smith's comparison of the robot to household pets suggests that the personal robot may be alive, but in the hierarchy of creatures it is not yet in the same league as the human species. The robot is considered a form of domestic animal: some type of space age pet.

Nolan Bushnell, inventor of the first video game, who founded Atari Computer Company and two personal robot companies, believes that home robots are ideal home companions. His creations include a cuddly, talking teddy bear and Petsters, plush electronic toy dogs and cats that come when called and lower their heads and tails and "go to sleep" when their human master turns off the lights. Bushnell claims that his "catsters" can do 70 percent of the things that real cats can do.

Doubters and skeptics point out that neither Bushnell's toys nor robots have consciousness. Intelligent machines, they say, do not "think" nor do they have independent motives. Instead, robots are much like typewriters or calculators. They do as they are told, as they are programmed. These critics admit the logical reasoning powers of computers and robots and their ability to retain and act from memory banks of stored data. But they contend that the robot lacks true intellectual prowess.

Among the human species, of course, we find those of high intellect, who think at a superior level, and we also know of men and women who do not—who seem to almost fit the definition of "robot." These people are predisposed to action rather than thought and frequently follow orders and instructions exactly and precisely, without hesitation or much forethought. Those who suggest that robots are alive have good cause to ask, "Are such people alive?"

Humankind has not yet fully resolved the question of what life is. Indeed, we have not even been able to classify many existing life forms to everyone's complete satisfaction. The theory of evolution remains controversial. Many scientists have come to question its tenets.

Some people believe that plants are conscious, that they can think and "feel" pain and are thus intelligent—if uncommunicative—life forms. In fact, we know that plants *can* communicate to us that they are thirsty, need fertilizer, too cold, too hot, etc. The achievement of biotechnologists in creating genetic engineered lifeforms further complicates our attempts to define life.

The Real Question

The real question that should be asked is not, "Are robots alive?" but rather, "What kind of life are they?" Current versions of robots lack conscious thinking skills and the ability to act of their own volition. Therefore, it may not be accurate to rate them on a par either with humanity or with animals. But in the near future, as technological breakthroughs bring us thinking computer brains for machines, we may well be forced to accept the fact that robots are roughly equivalent to humans in physical and mental performance. Then, people (and robots) may even call for robot liberation and equal rights.

Are Robots Male or Female?

Interestingly, the battle for women's rights seems already to be spreading to the robot world. When *Omni* magazine ran a feature story on robots, a controversy arose among the editorial staff over the type of artwork that should be used to illustrate the piece. Art director Elizabeth Woodson chose a painting of a female humanoid for "its beauty and for its synthesis of human and machine." But, writer Richard Wolkomir pointed out, "The robots in the article have male names, such as Isaac and Bob." (In the end the female humanoid was selected.)

The following year, when the U.S. House of Representatives passed a resolution designating a (male) robot as the symbol of National High Tech Week, Dee Snell-Wright of Baltimore fired off a letter in protest. Snell-Wright, you see, had created a female robot named Rebecca, and Rebecca was reportedly outraged that the Congress would fail to consider a female robot as the national symbol of technology. In response, Dr. Georgianna Anderson-Land, executive director of the House Science and Technology Caucus, explained that she was unaware that there was such a thing as a female robot. However, to mend the damage done, she promised to include Rebecca in some way. "I only hope there are no neutered robots out there," she sighed.

It is a fact that most android and personal robots are given masculine names: Maxx Steele, HERO, Buster, Rodney, and Marvin are among them. Perhaps this is because most robotics engineers and scientists, and most robot hobbyists, are themselves male. Still, the question of male or female is a thorny problem that so far defies resolution. One solution may be to treat all robots as unisex, or androgynous, as neither masculine nor feminine. The use of numbers, letters, and high tech monikers such as XR-III, R2D2, C-3PO, Sparky, Century, Cybot-1, Explorer, and Hubot denote the unisex nature of some robots, although traditionally masculine "personalities" are frequently ascribed even to many of these.

What Can Robots Do?

The more proper question here should be, "What can robots *not do?*" The range of robot behavior and performance has steadily expanded until, today, we have robotic machines that with their dexterity and brainpower, would astonish earlier human generations. They may not yet have the humor of Buck Rogers' wisecracking robot buddy, Twiki, nor be able to eat hot dogs as once did the lovable robot companion on the mid-'60s TV show, "Lost in Space." Nevertheless, today's robots are no mere buckets of bolts. They entertain, they delight, they save lives, and they do work that, for humans, is back-breaking or dangerous.

Appendix 2

The Anatomy of Robots

Present day robots are bodily and mentally quite different from you and me. For one thing, they are creatures of steel, plastic, silicon, and sometimes wood and rubber, while we humans have biological bodies based on carbon. Does this, then, render the robot inferior or provide *prima-facie* evidence that the robot is not alive? According to many experts the opposite is the case. While the modern robot has many limitations, its physical composition can often be an advantage.

In one Isaac Asimov story, "Reason," the robot automaton, Cutie, declares without contempt that humans are soft, flabby, and lack strength and endurance. Cutie points out that people depend for energy on the inefficient oxidation of organic material and that "the least variation in temperature, air pressure, humidity, or radiation intensity impairs your efficiency." The robot then asserts, "You are makeshift. I, on the other hand, am a finished product. I absorb electrical energy directly and utilize it with an almost 100 percent efficiency. I am composed of strong metal, am continuously conscious, and can stand extremes of environment easily."

Asimov's robots are *fictional* characters, we should hasten to mention, and what we want to do just now is examine the anatomy of real machines. Unlike Cutie, who had almost "100 percent efficiency" and was "continuously conscious," such is not the case for twentieth-century robots. Their limbs are often inefficient, lacking anything like the dexterity and the tactile (touch) sense of human beings. Only a few robots are mobile, and fewer still can climb stairs or navigate turns. The speech of many robots is human-like but on others, the speech is stilted, muffled, unclear, and occasionally inaudible.

Their vocabularies are often no larger than a few thousand words. Furthermore, an ophthalmologist would declare the average robot legally blind even though the more advanced artificial life machine can easily recognize faces, shapes and blurred edges, and can discriminate among colors.

The robot's brain is not nearly as sophisticated as that of a human. Of course, the microchip has given some robots the power to use memory and to calculate rapidly and correctly, but their creative abilities are deficient when compared to those of humans. Even the smartest machines have knowledge of only a few subjects. They may know how to paint an auto body or recite a story, but their repertoire is limited.

In contrast, even the average human being has a vast store of knowledge acquired from learning in a complex environment. The human has common sense, has insight into other humans' behavior, can "read" others' body gestures, and can even intuit events based on experience in dealing with the real world. The robot cannot do these things at all or else is severely underequipped in understanding and reacting to such complexities.

Still, the robot is a marvelous machine. The computerized robot has been with us less than the span of a single human lifetime. Robotics is in its infancy and yet we have hundreds of different robot machines, many with human appearance and an increasing number possessing rudimentary artificial intelligence and the five senses of *homo sapiens*. In some tasks their performance is superior to that of humans. The robot has bounded out of the science fiction realm and into reality. These machines are not yet able to live up to the potential of their kind, but their anatomy is already far advanced. It's exciting to peek inside and explore their anatomy.

The Robot Physique

The robot, a creature of a thousand shapes and forms, is configured according to its purpose. An assembly line robot may need only an arm to pick up and place items; therefore, a complete human form is unnecessary. However, there are many different sizes and types of "arm" robots. Robots used in electronics assembly are much smaller and lighter in weight than those used in factories to lift heavy parts or tons of unfabricated metal. More than one type of robot is built to resemble a snake; with a camera in its head it can do things such as travel down a pipe looking for seam cracks. Bomb disposal robots have squat, rugged bodies and manipulator arms and grippers to handle explosive devices. Similar manipulator arms are wielded by robots that handle radioactive materials.

Industrial robots can be outfitted with cutting torches, drills, lasers, sanding attachments, and other tools. They can also be equipped with vision to allow them to sort bad parts from good ones. Sensing a bad part, one robot machine can blow it into a trash receptacle with a powerful puff of air; another variety of robot simply picks up the part and discards it manually.

The Robot's Legs

Since we humans have two legs and are *bipedal*, we may expect robots to also have a pair, but engineers have found that that aspect of the human form is difficult to emulate because the control of human legs requires such highly complex coordination. In any case, the function of most robots simply doesn't call for two legs, so there's rarely any reason to construct them that way.

Roboticists Hirofumi Miura and Isao Shimayama of the University of Tokyo have built impressive bipedal robots, the BIPER-3 and BIPER-4, whose legs resemble stilts. The robots' legs have nearly the same degree forward flexibility as those of humans, but their knees also bend backward, They can walk forward with a pendulum motion and can also walk sideways.

In the believe-it-or-not category, at Carnegie-Mellon University in Pittsburgh, researcher Marc H. Raibert and associates have demonstrated a one-legged hopping machine that can hop in place, move straight ahead at a fast clip, or travel along a slightly more complex path, hopping all the way. This robotic pogo stick can be made to hop at different heights, too.

Researchers in the United States, Great Britain, France, and Japan have also built machines with three, four, five, and six legs and some of these robots are quite useful. They can navigate rugged terrain and step over obstacles. An example is Odex I, a unique robot built by Odetics, Inc., a California manufacturer mentioned previously. It is called a *functionoid* because of the many functions it can perform. With its spindly shape and six wiry legs, this robot resembles a spider. It can walk on rocky terrain, climb stairs, and carry loads of up to 2,000 pounds. Odex I may not be svelte and handsome, but he has a lot going for him. For one thing, he can stretch his height to 6 1/2 feet and shrink to 3 feet. When at his tallest, Odex I can squeeze through a narrow, 21-inch-wide aisle.

Odex I is remotely controlled by a computer and control unit and is equipped with sensing devices and a television camera. Its creators see the functionoid as a tool that allows man to explore areas never before accessible and to accomplish tasks that are either unsafe or impossible for humans.

Some of the fields Odex I will be exploring are forestry and fire fighting, coal and gold mining, law enforcement, and bomb disposal. The robot is able to enter a nuclear power plant after a dangerous incident, and, because he is waterproof, Odex I can function beneath the sea.

Unlike Odex I, many robots are either stationary or on wheels and rollers. Often the wheels or rollers are positioned to allow the robot to spin, to make quick turns, and to move forward, sideways, and in reverse.

The Robot's Arms and Hands

The arm of an industrial robot usually has three axes of motion (pitch, roll, yaw) and a wrist with the same axes of motion for a total of six axes. The hand—also called a gripper, claw, or end-effector—may have two or three parts ("fingers") or several. Three-fingered grippers are most common. The hand may also have affixed to it vacuum cups, magnets, hooks, or tools such as a spray gun or welding torch. The hand may even have a camera—an "eyeball"—in its palm. Robot arms and hands are controlled by the machine's brain, a built-in or remote computer.

Some roboticists stress that the robot hand need not duplicate the human hand because, for most tasks, a five-fingered hand is not necessary. They believe that a hand should conform to the purpose for which it is to be used. Nevertheless, inventors, engineers, and researchers continue to work at developing a robot hand that is as dexterous and useful as the human hand. While a three-fingered gripper may suffice for 90 percent of all industrial tasks, the remaining 10 percent of tasks represent a huge deficit. The greater the capability of the hand—the closer it approximates the human appendage—the more tasks it can perform, making more applications possible.

Inventor-engineer David Petersen, founder of Victory Enterprises Technology in Austin, Texas, has invented a robot hand that may be about as close to a human model as now exists. A 35-year-old former project engineer for Houston Instruments, Petersen says that "the biological systems of the human hand have much to teach us, and an understanding of these systems would be useful even if the final system departed substantially from the biological system."

Petersen's robot hand uses three fingers and a thumb and is incredibly nimble and efficient. It can, for example, grasp an electric drill, a ballpoint pen, or a coffee cup. Unlike other robot hands, this one has sixteen axes, or motions. One motor controls all sixteen hand joints. Petersen is working to add vision to his hand system so it can see and feel.

Other researchers are also "lending robots a hand." Dr: John Purbrick of M.I.T.'s Robotics Laboratory is busily creating artificial skin and materials to enhance the robot's sense of touch. And in Tokyo, a team at the Electromechanica Laboratory has been able to create a three-fingered hand that can twirl a baton.

The Robot's Muscle and Nervous System

The human body has about 300 muscles capable of as many as 1,000 motions. Nerve tissue interconnects throughout the body, and both muscles and nerves are controlled by the brain. To perfectly duplicate this, a robot would need 300 electric motors; it would also require incredibly complex electronics and electrical systems wired via integrated circuitry and fiberoptics into a ganglion system. The system would be controlled by a sophisticated computer brain. Since such a complex system hasn't yet been built, current robots can duplicate only a few of the capabilities of their flesh and blood counterparts.

Robots may be powered by electricity, hydraulics (pressurized liquids, usually oil) or pneumatics (compressed air). Electrical power is used only for smaller robots, such as home models, because more powerful models would require huge, overly bulky electric motors. Small home robots often have batteries on board. Several intelligent home robots are able to sense when these batteries need recharging, and they then go to the nearest electric outlet and plug themselves in. Hydraulics is the preferred system for most industrial robots.

Intelligent robots have a central processing unit, or computer, which, after programming, transmits signals—bits and bytes of information in computerese language—to an interface. The interface has an electronic switch for each of the robot's motors, and the motors control the movement of "muscles" or joints. The motors do this by working gears attached to shafts.

The Robot's Skin

The robot's torso, the main part of its body, usually is made of steel. However, the skin of most personal robots is fashioned from molded plastic, and a few hobbyists have used wood for their robot's body casing. The body of show robots is often a combination of steel and plastic. Those building androids are now using material that enable the robot to make facial gestures.

The Robot's Brain

Robots have been lurking around the world for several decades now; but the usefulness of the early, hulking, electromechanical clunkers was limited to a few heavy industrial tasks. Then the microchip came along and, suddenly, the relatively witless robot had an enhanced brain. The new silicon brain hastened the evolutionary process for the new life form and, within the space of a few short years, the robot has taken on a wholly new physiological configuration.

Today, worker robots are replacing thousands of humans in the workforce, and some personal robots have "personality" and even charisma. The IQ of robots is growing by leaps and bounds. Thanks to the microchip, robots are becoming "near people." The microchip is the foundation of the robot's store of knowledge and the source of its capabilities. Until the advent of microcomputers, the worker robot was nothing more than another machine tool, and personal robots were nothing more than radio-controlled hulks.

Members of the first generation of industrial robots we're able to perform only single tasks, but today's improved robots possess fully programmable computer brains that can be reprogrammed again and again, allowing these machines to switch from one task to another. This makes them much more versatile and useful.

Some robots have a degree of artificial intelligence that makes it possible for them to fully understand a number of problems and to figure out the best way to accomplish their tasks. Many of these robots can "see;" they are able to hear commands spoken by humans and to act as ordered. If something goes wrong on an assembly line, they autonomously detect it, stop the line, and signal human attendants to come and check. At IBM, one such two-armed robot has a job as an electronics inspector. It does the work of twenty people, checking wired connections to make sure they are soldered correctly.

Programming a Robot's Brain

To program the industrial robot's computer requires knowledge of sophisticated computer languages. Industrial robots can be programmed from a hand-held teaching pendant or from a remote computer console keyboard. They can then also be tested and put through their simulated paces on a CAD/CAM terminal before they're actually started on their new task and movements. Preprogrammed software systems are available to help speed the process.

The Robot's Sensory Faculties

The five human senses—taste, touch, smell, sight and hearing—have been difficult to duplicate in robots. But, astounding progress has been made. Let's examine the sensory equipment, or faculties, of electromechanical life.

Tactile Sense (Touch)

Some robots incorporate a wide range of tactile, or touch, sensors. At the University of Wollongong in Australia, Professor R. A. Russell has supervised students in the development of a robot electronic thermal sensor that, when installed in the robot's hand(s), allows the machine to feel heat or cold. When the robot picks up a block of metal that is too cold or too hot, it drops the block and returns to its original position.

There are many other examples of the robot's fine sense of touch. Hitachi has built a pressure-sensitive hand with silicon rubber insulators that can grip and manipulate an egg without cracking it. The Robot Factory, a maker of show and promotional robots, offers a robot that can whip out a business card and present it with its gripper hand.

Olfactory Sense (Smell)

Yes, some robots have a nose. It may not look like your nose or mine, but it does give the robot the ability to sniff out and smell odors. In Great Britain, scientists have developed a robotic olfactory receptor that can detect and distinguish the smell of a rose. Researchers at Carnegie-Mellon's Robotics Institute are developing integrated circuitry materials on a microchip that are sensitive to several types of gases. When installed on robots, these microchip sensors enable the machines to detect dangerous, toxic gases and chemicals in the environment. Some industrial robots are able to smell a poisonous gas or vapor emission and sound an audible alarm or send an electronic signal to warn humans.

There are robots that can "smell" and detect smoke or sense heat and act as fire alarms. The RB5X personal robot, for example, can be equipped with an optional fire protection system. When it senses a fire, it calls out "Fire, fire, fire," then goes into action with an on-board fire extinguisher. Also, many hobbyists have mounted smoke detectors on their home-built or commercially bought robots.

Vision

Artificial intelligence scientists and engineers are making a great deal of headway in the field of robotic vision. When shown various photographs, robotic eyes can distinguish between such shapes as that of a woman or man, elephant or buffalo. Facial recognition systems are in use. At the University of Rhode Island, a team created a robot system that uses vision to locate and pick up small, randomly oriented objects in a work bin, determine their shape and select the correct ones as ordered by its computer program. Japanese researchers at Mitsubishi and Hitachi have also developed vision systems that can do the same thing. And at the University of Hull in Great Britain, a robot system can recognize letters of the alphabet.

Vision robots are terrific additions to assembly lines. The Bulova Company uses robot watchmakers to assemble time pieces. At Honeywell in Minneapolis, Minnesota, robots inspect solder joints on printed circuit boards. The robots inspect 1,500 spots per minute, searching for bridges, pits, and missing solder. At a Ford Motor Company plant in Wayne, Michigan, a "picky" robot equipped with photoelectric cells for vision, scans auto instrument panels of four different colors and selects the correct one needed to match each automobile being assembled.

There are several types of robot vision systems. One uses ultrasonic sound waves to locate and identify an object. Lasers also are used both to help robots know where they are and to calculate the dimensions of objects. Some robots have laser holographic scanners that read bar codes just as do modern supermarket check-out scanners.

However, the most common vision system is a television camera mounted on the robot. The camera translates what is seen into picture elements, called pixels. Pixels are given numerical values based on the varying levels of light, and the robot's computer brain compares these values with those stored in its memory bank. The robot already has images stored in its bank and compares what is seen with these stored images. New microcomputer silicon chips are being designed to enhance the recognition process. Another innovation makes use of a fiber optic sensor installed directly in the robot's hands. These fiber lines are able to obtain better quality pictures and send these on to the camera.

Today we have robots whose vision is equal or even superior to that of humans. Some robots are equipped with vision devices sensitive to infrared. ultraviolet, and other electromagnetic wavelengths beyond human discernment capacity.

Speech

Robots speak but their voices often sound machine-like, as though they have a speech impediment. Also, their vocabulary is quite limited. Most robots that speak are of the personal, or home, variety. Speech synthesizers can be added or are built-in. Most have sixty-four phonemes, or sounds, that can be used to create different inflections and amplitudes. With a synthesizer, robots can sing songs and recite poetry. With a modem, they can answer your telephone. In one noted experiment, an RB5X personal robot in Colorado called a TOPO robot in California for a telephone chat. The iPhone has "Siri" capability to answer your questions and perform receptionist tasks.

AT&T is working on some fantastic speech synthesizers available for robots, One gives speech quality approximating that of humans. Eventually all the words in a very large dictionary will be maintained on a laser disc or a microchip in the robot along with instructions for grouping these words into recognizable phrases.

In all speech synthesis systems that yield a natural-sounding voice, a vocabulary of key words and phrases is spoken into a microphone, digitally encoded and stored in the robot's computer memory. The robot then draws upon that encoded vocabulary to put together meaningful sentences. One can also direct certain types of robots to speak by typing a message on an attached or remote computer keyboard.

Hearing (Auditory)

More and more robots are being given "ears." For example, Kearney and Trecker Corporation of Wisconsin has a system in which a human operator can give spoken commands to an industrial robot by using a hand-held radio transmitter. Once the robot has done what it was told to do, it sends back an oral message to its human master. Arctec Systems, a Columbia, Maryland, company produces and markets a system that lets you talk to your robot. With Arctec's Micro-Ear system, you simply speak into the microphone and the robot digests what you say and remembers words with about 98 percent accuracy.

A Brief Glossary of Robotics Terms

Below is a brief glossary of commonly used robotics terms.

Android. A robot that closely resembles a human being in physical appearance.

Artificial Intelligence. Man-created machine intelligence comprised of human intelligence functions such as induction, deduction, reasoning, and adaptation.

Automaton. A mechanical apparatus, usually in the form of a human, animal, or bird, that functions repetitively and cannot be reprogrammed.

Central Processing Unit (CPU). The control center of a computer in which one or more microprocessors direct the robot's sequence of operations and executions.

Cyborg. CYBernetic ORGanism. Also called a bionic man or woman. A human modified by the addition of artificial (robotic) body parts.

End-Effector. The hand, manipulator, or gripper, of a robot.

Hardware. The physical apparatus of a robot.

Humanoid. An android robot that closely parallels the form of a human in most essential features.

Language. A set of symbols and rules for robot communication and operation.

LED. Light emitting diode characters that provide an illuminated visual display.

Mechanoid. An android robot roughly approximating the human figure but also with at least some essential features of a machine or mechanical device.

Microprocessor. Integrated circuitry that directs computer (and robot) operations.

Pick-and-Place Robot. An early generation arm robot limited in function to picking up and placing items point-to-point.

Robotic. Pertaining to, or about, robots.

Sensor. An input/output feedback device that permits the robot to interact with its environment.

Software. The computer program of prescribed and ordered sequences that controls and directs the functioning of a robot.

Index

Symbols
7th Indian Robot Olympiad 103
2001: A Space Odyssey (movie) 120
2005 World Expo in Japan 14

A

Ackland, Nigel 36
acoustics 148
Activistpost.com 58
actroid 55, 57
Adam and Eve 68
Advanced Robotics 170
aerio-nautical man 75
Aftergood, Steve 53
Age of Electricity and Machines 132
Age of Spiritual Machines, The (book) 26, 27
Aguilera, Christine 150
Aiken, Howard 137
air-launched-cruise-missile 192
Aizawa, Jiro 92
Albo (robot dog) 61
Albus, James S. 220
alchemy 16, 17
Alchemy of Change, (book) 11
Alchemy of Stone, The (book) 116
ALCM 194
aluminum man (automaton) 134
Amazing Stories (magazine) 106, 108
Anderson-Land, Dr. Georgianna 222
Androbot, Inc. 140, 162, 163
android 33, 48, 57, 61, 65, 149, 218, 219, 221, 222
Android Amusement 179
Android (movie) 118
Android and The Human, The (speech) 62
Android Open (software) 11
Androman (robot) 163, 164
androwagon 162
Andy (robot) 163
Anne Arundel Community College 180, 181
Antichrist 43, 44
Aphrodite 68
Apocalypse 21, 30
appear to cry 14
Aquinas, Thomas 69
Archimedes 130, 140

Arctec Systems 229
Are Computers Alive? (book) 97
Are robots alive? 220
Aristotle 67
Armadillo (robot) 164
Armageddon conflict 48
Armatron (robot) 165
artificial arm 35
artificial arteries, ears, eyes, heart, joints, bones, tendons, ligaments, kidney, larynx, nerves, nose, spinal cord, veins 142-145
artificial intelligence 14, 15, 23, 24, 47, 48
Artificial Intelligence (movie) 33
Asada, Minoru 62
Ashworth, Diane 144
Asimov, Isaac 105, 106, 110, 111, 112, 129, 169, 205, 213, 223
asp (robot) 131
Astounding Science Fiction (magazine) 106
Astounding Stories (magazine) 106
Atomic Age 109
audioanimatronic devices 220
Autobots 84
Automan (TV series) 122, 148
automation 23
Automatix 174
automaton 69, 72, 74, 77, 218, 219, 220
autonomous automobile 150
autonomous robots 29
autonomous weapons 30
avatar 43, 44, 46, 47, 51, 53, 54
Avatar (movie) 44
Avengers, The (TV series) 122
Axlon Corporation 163, 166

B

Babbitt, Spencer 133
babysitter (robot) 125
Bacon, Roger 131
Bailey, Ronald 33
Bakaleinikoff, Bill 181
Bandai 84, 85, 87
Bannadonna 77
Barnett, J.M. 134
Basic Telecommunications Corp 151

Battelle Memorial Institute 88, 89, 211
Battlestar Galactica (TV show) 122
Baum, L. Frank 80, 82
Baxter (robot) 176
Beall, Sylvia 97
Beley, Gene 179
Bell Tower, The (short story) 77
Bendix 194
ben Judah, Rabbi Elijah 71
Bicentennial Man (movie) 33
Big Brother—The Orwellian Nightmare Come True, book 24
Billina (the chicken) 81
Bill of Rights 30
Binder, Eando 106, 109
bio-chemistry 38
biochip 9, 146, 152
biodesigners 40
Bioengineering 38
biomechanics 151
biomimetics 141
bio-morphic machine 146
bionic human 141, 220
bionics 33, 141, 142, 145, 146, 148, 151, 152
Bionic Woman, The (TV series) 149, 220
bioplastic 145
bioroids 48
biosensors 145, 146, 152
bipedal 224
Black Hole, The (movie) 120
Blade Runner (book) 48, 64, 114, 115
Boeing 194
Bolingbroke, Ken 143
Bomb Disposal Technician 177
Bonds, Barry 37
Bondurant, Phil 89
Bonham, Doug 205
Borden's Dairies 179
Borg (of *Star Trek*) 26
Bosworth, Joe 129
Bozo (robot) 128
Bracelli, Giovanni Battista 69
Brain Mind Institute 26
Brave New World (book) 48
Brazil 21, 22
Bristol-Meyers 146

Brooks, Dr. Rodney 15, 26, 27, 80, 90, 176
Buchwald, Art 98
Bushnell, Nolan 129, 140, 163, 221
Business Week (magazine) 210
Butler, Samuel 75, 105

C

C-3PO (robot) 118, 119, 153, 217, 219, 222
Caan, James 181
Calvin, Dr. Susan 110
Cambridge University 48, 49
Cameron, James 44
Campbell, John W. 105, 106
Capek, Karel 77, 79, 217
Captain Future (series) 105
Carbomedics 146
Careers With Robots (book) 140
Carnegie-Mellon University 90, 206, 212, 224
cartilage induction factor 145
Cassandra (android) 119
Caves of Steel, The (book) 112
CB2 (child robot) 63
Center for Intelligent Machinery 91
Center for the Study of Existential Risk 48
Center of Neuroscience and Technology 26
Central Intelligence Agency (CIA) 20, 30
Centre for Existential Risk 49
Century I (robot) 87, 88
Cheetah (robot) 45, 196
Chicago World's Fair 97
Chrysler Corporation 172
CIM 172, 174
Cincinnati Milacron 90
Cisco 21, 25
civilian casualties 20
Clark, Barney 145
Clarke, Arthur C. 120
cloning 40
cochlear implant 142
Cokebot (robot) 190
Collagen Corporation 146
Colossus (book) 115
Coming Technological Singularity, The (book) 21
Coming Technological Singularity, The, (paper) 24
Commander Cody 120
Commander Robot (robot) 182
Compubot (robot) 167
computer-integrated-manufacturing 172
ComRo 88
Comro Tot 160, 161
Condon, Bernard 58
Conquest of the Space Sea (novel) 128

Consolidated Diesel Corporation 140
Cooley, Dr. Denton 145
Cooper, Pat 53
Corbin, Charles 115
Cornish, Edward 41
Council on Foreign Relations 33
crab robot 104
Cretectos, George 206
Crowley, James L. 206
Cubot 88, 89
Cummings, Ray 106
Curiosity Rover (robot) 199
Cutie (robot) 223
Cyberiad, The (book) 113
Cyber I (mechanical man) 138
cyberman 220
Cybernetics: Control of Communication in Animal and Machine (book) 139
Cybernetic Walking Machine 212
cybert 220
cyborg 28, 33, 40, 220
cybot 220
Cy-Kill 84, 87
Cyro 2000 170

D

Dacron grafts 146
Daedalus 68
dancing robots 19
Darke Hieroglyphics: Alchemy in English Literature (book) 17
DARPA 17, 29, 35, 147, 196,
Darwin Among the Machines (essay) 75
da Vinci, Leonardo 131
Davis, Nan 151
Dawkins, Richard 41
dawn of robotics 129
Day the Earth Stood Still, The (movie) 118
DC-2 (robot) 179, 181
de Camp, L. Sprague 112
de Chardin, Pierre Teilhard 51
Defenders, The (book) 115
Defense News (publication) 53
de 'Isle-Adam, Villiers 105
Dell 21
del Rey, Lester 106, 110
Demon Seed (movie) 120
Denning Mobile Robotics 88
Department of Defense 18
depression 23
Descartes, René 69
Descendance (play) 58
designer babies 40
d'Esnon, A. G. 209
de Vaucanson, Jacques 131
Devol, George C. 140
DeVries William 145
Dextre (megabot robot) 200, 201
Dice, Mark 24
Dick, Philip K. 48, 62, 64, 65, 114, 115

Diego-san (robot) 66
Dingbot (robot) 87, 159
DNZ sniffers 18
Do Androids Dream of Electric Sheep? (book) 64, 114, 115
Doberman robots 205
Draganflyer X6 (drone) 192, 194
Draughtsman, The (automaton) 131
dream machines 51
Droidbug 154
drone 17, 19, 20, 191, 192, 194, 198, 218
drone aircraft 29
drone technology 20
duck (robot) 131
Dyer, Bob 89
Dynaman 85
dystopia 23
dystopian "Utopia" 48

E

EATR 18
EcoRP (painting robot) 102
Thomas A. Edison 105, 133, 134, 140
Egyptians (ancient) 67
Einstein, Albert 98
Elami, Jr. (robot) 161, 163
Electromechanica Laboratory 226
electromechanical warriors 191
Elektro (mechanical man) 134, 136, 137
Elijah Bailey (robot detective) 112
Ellis, Edward S. 75
emotion 14
empathy 14
end-effectors 171
Energetically Autonomous Tactical Robot 18
Engelberger, Joseph F. 140, 169, 204, 206
Epic of the Creation (story) 68
Epimetheus 67
Eraser (movie) 33
Erewhon (book) 75
Estes, Vern 98
eugenics 40
Evans, Oliver 133
Ever-1 (robot) 57

F

facial recognition cameras 18
Factory of the Future 172
Fantastic Adventures (magazine) 106
Fantasy-Thrilling Science Fiction (magazine) 106
Faraday, Michael 132
Federation of American Scientists 53
feeling 13
Ferraro, Geraldine 96
fiberoptics 142
Fire Fighter 177

Flesh and Machines: How Robots will Change Us (book) 27, 80
Flight Instructor 175
Florida Robotics 186, 187
Florida State University 91
FM-2030 34
Forbidden Planet, The (book) 117
Forino, Michael 158
Foxbots (robot) 22
Foxconn 21, 22, 58
Frankenstein 19, 71, 72, 73, 77, 119, 120, 122
F.R.E.D. (robot) 162
Frude, Neil 206, 212
Fujitsu Fanuc factory 91
Fukuyama, Francis 33
functioning replica of a human brain 26
functionoid 177, 219, 225
Future Combat Systems 18
Future Eve, The (book) 105
Futurist, The (magazine) 41

G

Galaxy Science Fiction (magazine) 35, 114
Gardian (robot) 87
Gasman, Lawrence D. 205
GCA Corporation 206
Gemini (robot) 161
Geminoid-F (robot) 42
General Dynamics 91, 194
General Electric 98, 169, 171, 211
General Motors 140, 169, 172
Genetic Engineering 40
Gernsbeck, Hugo 105, 106, 107
Gibson, William 116
Global Future 2045 World Congress 42
Glover, Jonathan 51, 53
GM Fanuc Robotics 171
GoBot (robot) 84
Godaikin (robot) 85
Gog (robot) 119
Golden Age of robots 122
Golden Age of Science Fiction 106
golem 71
Golion (robot) 85, 87
Gort (robot) 118
Gou, Terry 21
Great Work 16, 17
Greece (classical) 67
Guardian, The newspaper 20

H

Hackwood, Dr. Susan 146
Hadaly (android) 105
Hairy Three Piece Band (robots) 183
HAL (robot) 120
Hamilton, Edmond 105
Handbook of Robotics (book) 110
Handy Corp. 19
Hankins, Walter 213
Hanson, Dr. David 65

Hanson Robotics 65, 66
Hard Drive (robot) 186, 187
Harrison, Harry 116
Hasbro-Bradley 84, 86
hearing 13
Heathkit Company 96, 140, 155, 205
Heimes portable heart driver 145
Heinlein, Robert 112
Heinz 179
Helen O'Loy (robot) 110
Helmers, Carl 161
H.E.N.R.Y. (mechanical man) 139
Hephaestus 67, 68
HERO 1 (robot) 88, 96, 140, 155, 222
Hero, Jr. 155, 156
He, She, It (book) 116
Hewlett-Packard 21
high radiation areas 90
Hitachi 93, 227, 228
Hoffman, E.T.A. 105
Hollow My Weenie (movie) 119
Holmes and YoYo (TV series) 122
home robots 153
homo sapiens 25, 67, 203, 224
Hon Hai Precision Industry 21
Hoover, J. Edgar 98
Ho, Professor 104
Horror Machine 53
How Life Learned To Live (book) 148
How Mind Controlled Robots Work (lecture) 101
How to Be a Real Man (book) 41
How to Create A Human Brain (book) 47
HRP-4C (android robot) 61
Hubot (robot) 158, 160
Hubotics 158
Huey, Dewey, and Louie (robots) 219
Hughson, J. D. 134
human and robot romance 57
human-animal humerics 48
Human clones 48
human era will be ended 24
human intelligence 14
Humanist Religion 34
humanoid 13, 33, 218, 219, 222
Human Rights Watch 30
human-robot interactions 64
humans become cyborgs 141
humans into robots 32
human underclass 27
Hunt, LtCol Joe 196
Huxley, Sir Julian 34, 35, 48
hydraulics 226
Hyperion Cantos (book) 116

I

Ideal Toy Co. 87
IEEE Spectrum (publication) 196
Immortalism 38
immortality 16, 17

Impact—Recession, Tech Kill Jobs (article series) 58
Industrial Revolution 23, 25, 75
InfoWars (magazine) 43
infrared 161
Institute of Automation of Chinese Academy of Sciences 56
Intel Corp. 13, 21
Intelligent Robot Research Center (Korea) 57
International Committee for Robot Arms Control 19
International Computers in Engineering Conference and Exhibit 209
International Flexible Automation Center 90
International Robotics 184, 185, 188, 189, 190
invasion of the robots 215
iPhones 16, 27, 58
I, ROBOT (book) 110
Iron Jaw (cartoon character) 34
Iron Man (movie) 121
Iron Terror (book) 106
Ishiguro, Professor Hiroshi 14, 62, 63
i-Sobot (robot) 61, 159
Dmitry Itskov 42

J

Jacquet-Droz brothers 72, 131, 132
Jarvik, Robert 145
Jarvis, Eugene 213
Jetsons, The (TV series) 122
Jewish Wonders (tract) 71
Jones, Alex 43, 45
Jones, D. F. 115
Jong-Hwan, Kim 57
Joy, Bill 25, 26, 45
Jung, Carl J. 16

K

Kabbalistic magic 71
Kanamits 18, 19
Kato, Ichiro 91
Kawada (robot) 61
Kearney and Trecker Corporation 229
Kermit the Frog (robot) 86
Kilby, Jack 139
K.I.R.K. (robot) 186
K.I.T.T. (robot) 122
Knight Rider (TV series) 122
Kokoro Robot Corporation 55, 66
Kolff, Willem 145
Krull, Ulrich 145
Kuratas (robot) 16, 196, 198
Kurzweil, Ray 15, 26, 27, 28, 38, 42, 45, 47, 48, 91

L

Lady Musician, The (automaton) 131, 132
Lang, Fritz 116, 117

Langley Research Center 213
Laproscopic surgery robot 216
larynx 145
Lee, Tanith 115
Legos 207
Leifor, Dr. Larry 151
Lem, Stanislaw 113
Lescho, Jean-Frederic 131
lethal technocracy 20
Levy, David 26, 206
Lindbergh, Charles 141
Lind, Dr. Michael 88, 90
Link, Adam 108, 109
lion (robot) 131
Lisack, Professor J. P. 209
Living Brain, The (book) 139
Lockheed 194
locomotion 148
London Magazine, The 105, 106
Looking Out For Number 1 (book) 41
Lost in Space (TV series) 122, 222
Love and Sex With Robots (book) 26, 206
LS3 AlphaDog (drone) 196, 198
Lucas, George 105, 118
Lukyanova, Valeria 38
Luna 17 (robot craft) 199
Lunakhod (space rover) 200

M

Machine Age 75
machine guns 18
Machine Intelligence Corporation 174
Machines to Conquer 191
Mackie, Harry 206
MacLeod, Ken 116
Ma fille Francine (automaton) 69
Magnus, Albertus 69, 131
Maillardet, Henri 132
Malone, Robert 212
Malthus' notion 29
Mälzel, Johann 72, 75
Marduk 68
Marilyn Monroe (robot) 93
Markram, Harry 26
Martin-Marietta Corporation 209
Marvel Comics 86
Marvin (mechanical man) 137
Minsky, Marvin 213
Mason, Carl 151
Massachusetts Institute of Technology 80, 90
Master Slave 42
match.com 57
Matrix (movie series) 33
Max 404 (robot) 118
Maxx Steele (robot) 84, 87, 162, 163, 222
McGuire, Paul 48, 51
McKinsey Global Institute 58
Mechanical Engineering (magazine) 148

Mechanical Man, The (movie) 128
mechanoid 219
medical aides 176
Medieval alchemists 17
Mega Forces: Signs and Wonders of the Coming Chaos (book) 34
Melville, Herman 77, 105
Memo Robot (robot) 168
Metropolis (movie) 58, 116
Micro Air Vehicle (drone) 198
microchip 137
Microsoft 21
Military Robots 191, 195
Miller, Ritz 179
Mindstorms EV3 (robot) 207
mingle.com 57
Ministry of Manufacturing Trade and Industry (Japan) 206
Minksy, Marvin 90, 204
MIT University 13, 15, 26, 80, 90
Mitchell, Claudia 35
Mitsubishi 228
Miura, Hirofumi 224
Monsanto 146
Moore, George 133
Moore, Gordon E. 13
Moore's Law 13, 15
Moravec, Hans 26, 90
More, Max 34
Moskal, Joseph 152
Motorola 21
Movellan, Javier 66
Mowbray, Melton 90
Mr. Atomic (robot) 93
Mr. Ichiro (robot) 92
Mr. Juro (robot) 93
Mr. Kuro (robot) 93
Multi-Robot Pursuit System 18
Muppet (magazine) 86
My Living Doll (TV series) 122
mythology 16

N

Naked Sun, The (book) 112
nanorobotic implant 28
nanorobots 47
Nanotechnology 146
NASA 199, 200, 201, 202
National Bureau of Standards 220
National Geographic Society 177
National Institute of Health 152
Naval Research Lab Attempts to Meld Neurons and Chips (article) 53
Naval Research Laboratory 53
Neal, Patricia 118
Nearing Singularity, book 26
Neato VX21 (robot) 165
Neph 68
neural chips in the brain 53
Neural Computations Machine Perception Laboratory 66
neural implant 28

neural road-map 47
neural technology implants 28
Neuromancer (book) 116
New Republic, The (magazine) 36
New Scientist (magazine) 18
newswithviews.com 48
New York Times 18
New York World's Fair 136
Nexi (robot) 13
NextFest 56
Nielsen, Niels E. 115
Night Sessions, The (book) 116
noosphere 51
Nuclear Plant Inspectors 175
nuclear power plants 90
Nussbaum, Bruce 210

O

Obama, President 36
Odetics, Inc. 177, 210, 219, 225
Odex I (robot) 177, 225
Ohio State University 91
Oleson, Douglas 211
Olympia (robot) 105
Omnibot 159, 160
Omni (magazine) 222
Onaga, Eimei 61
optics 148
organ regeneration 146
Orlando, Nancy 213
Osaka University 14, 61, 62, 63
Our Final Hour (book) 48
Ovid (Roman writer) 68
Ozma of Oz (book) 81

P

paraplegics 151
Pare, Ambroise 141
parity with human brains 23
Pentagon 18, 29, 191, 195
perfection 16, 17
personal robot 153, 154, 155
Personal Robot Book, The (book) 129, 130, 140
Personal Robotics Corporation 96
Personal Robot Market 61
Peterson, David 208, 225
Petraeus, General David 30
Petrofsky, Dr. Jerold 151
Petsters (robots) 221
Phantom Empire (movie) 120
piano-playing robot 19
Piercy, Marge 116
Piotrowski, George 148
Pistorius, Oscar 37
Plato 67
pneumatics 226
Poe, Edgar Allan 74
Policeman's Beard Is Half-Constructed, The (book) 213
Pollard, Bernard 36
polyethylene grafts 146

population control 28
positive genetic engineering 41
Postgenderism 38
Predator UCAV (drone) 192, 214
Price, Huw 48
pro-football 36
Project L.U.C.I.D. (book) 9
Prometheus 67, 72
promise of immortality through robots 42
Psycho-Brain 47
psychopath 47
PUMA (robot) 178
Purdue University 210

Q

QRIO (robot) 64
Quadracon (robot) 181
quadriplegics 151
Quasar Industries 88

R

R2D2 (robot) 118, 119, 153, 217, 219, 222
Radio-controlled robots 179
Raibert, Marc H. 224
Rainbows End (book) 116
RB5X 88, 155, 157, 158, 221, 229
RB Robot Corporation 129, 221
R. Daneel Olivaw (robot) 112
Reade, Frank 77
Rebecca Robot (robot) 96
Recollection of Six Days Journey in the Moon (article) 75
Rees, Lord Martin 48, 49, 51
Religion Without Revelation (book) 34
Rennie, Michael 118
replicator robot 26
Research Institute of America, The 206
Rethink Robotics 176
Return to Oz (book) 82
Revelation (book of the Bible) 43, 44
Ripley's Believe It Or Not (TV show) 171
Riptide (TV series) 122
Rise of the Robots—The End of Humanity? (article) 43, 45
Robbie (robot) 118
Robocop (1924) 127
RoboForce 84, 87
Robonaut (robot) 200, 202
Robonosis.com 185
robota 217
Robot Alley 91
robot arm wheelchair 59
Robot Book, The (book) 212
Robot Factory, The 182, 183, 184, 227
Robotic Industries Association 209
robotic planes 19
robotics age 96, 209
Robotics Age (magazine) 161

Robotics Industries Association 129, 218
Robotics Institute 90
Robotic wheelchairs 151
Robot Invasion 83
Robotland 206
Robotman (cartoon) 84
Robot Monster (movie) 120
robot physique 224
Robot Redford (robot) 181, 182
robot revolution 55, 59, 61, 137
Robotron (video game) 213
robots as sexual partners 26
robots as warriors 29
Robot Shop, The 154, 165, 167
Robots (movie) 121
Robots of Dawn (book) 112
robots of science fiction pulps 106
Robot/X News (publication) 178
Roboz (robot) 122
rocket launchers 18
Romaster, Major Clyde 192
Rong Cheng (robot) 56
rooster (robot) 131
RoPet-HR (robot) 96
Rosenblatt, Gideon 11
Rosen, Charles 140, 174
Rossum's Universal Robots (R.U.R.) (play) 58, 77, 79, 217
Rotwang 117
Royal Knight (automaton) 134
Rubik's Cube 88, 89
Rucker, Rudy 116
Rulers, The (book) 115
Rules, The (book) 41
Russell, Professor R. A. 227

S

safety of humans who work alongside robots 97
Saffo, Dr. Paul 9, 11, 17, 18
San Diego State University 24
Sandman, The (book) 105
Satan's Satellites (movie) 120
Saturn's Children (book) 116
Scarab Robotics Corporation 175
Scarab X-1 (robot) 175
Schaeffer, Roland 97
Shockley, William 137
Schuder, Johann Jakob 71
Schwarzenegger, Arnold 30
Science Museum, Tokyo 93
Science News (magazine) 53
Scribe, The (automaton) 131, 132
Second Genesis 34
Security Robots 87
Sedia Ekaterina 116
seeing 13
Segway HT 200
Selfish Gene, The (book) 41
selfish replicators 41
self replicating robots 22

sensory channels 53
Servitron 154
Shadowhawk (drone) 192
Shakey (robot) 140
Sharkey, Dr. Noel 19, 30, 194
Sharp 21
Sheffield University 20, 30
Shelley, Mary 72, 73
Sheppard, H.J. 17
Shimayama, Isao 224
SICO 185, 188, 189
Side Impact Dummy (android) 219
Silent Running (movie) 219
Silicon Gods (book) 53
Silicon Valley 90, 91
Silver Metal Lover, The (book) 115
Simak, Clifford D. 105, 106
Simmons, Dan 116
Sim One (robot) 219
Simons, Geoff 97, 221
Singularity Hub (publication) 21
Six Million Dollar Man, The (TV series) 33, 142, 220
Six T. Robot 184
Skywalker Luke 118
Sladek, John 115
Small Wonder (TV series) 122
smartphone technology with robotics 147
smelling 13
Smith E. E. 106
Smith, Will 111
Snail (robot) 177
Snell-Wright, Dee 96, 222
social relationships and robots 55
Society of Manufacturing Engineers 129
sociology of robotics 212
sonar 155
Sony 19
soul 14, 17
SoundWave (robot) 84
soylent green 28
Space Shuttle 200
Sparko (mechanical dog) 134, 136
Spencer, Christopher 133
SRI International 140
stair-climbing robots 19
Stanford University 9, 11, 13, 17, 151
Star Trek (TV show) 122
Star Wars (movie saga) 105, 118, 119, 217, 219
Steam Driven Boy, The (book) 115
Steam Man of the Prairie, The (book) 77
Stelarc 54
Stepford Wives, The (movie) 26, 120
Stine, G. Harry 53
"Stop the Killer Robot" campaign 30
string-controlled puppets (robot) 130
Stross, Charles 116
Sun Microsystems 25

superbeings 36
superhuman abilities 33
superhuman intelligence 21, 24, 34
Superior Robotics 181
Supermarket Assistant 177
Super Race 40
surgeon 178
surveillance drones 20
Sutherland, Ivan 212
Symbion 142
Symbolics Computer 90
synthetic man 67, 68, 71

T

tactile 155
Talkabot (robot) 166
Talking Doll (automaton) 133
Tallinn, Jaan 48
Talus 67
tasting 13
Taylor, Bruce C. 139
technological singularity 23, 24, 38
Techogeanism 38
teleoperator devices 218
telephone lineman 176
telepresence 42
Televox (mechanical man) 134
telos 51
Tenbo R-1 and R-2 (robots) 174
Terminator (movies) 30, 33, 44
Tesar, Delbert 90
Tesla, Nikola 134, 135
Texas Instruments 139
thermodynamics 148
thinking capacity 15
Thinking Machines, Inc. 90
Thomas' Gospel of the Infancy of Jesus Christ 68
Thompson, Michael 145
Thomson, Amy 116
Three Laws of Robotics 110
Tik-Tok of Oz (book) 81, 82, 115
Tin Man 80, 81
Titanoboa (robot) 196, 197
Tolles, William 53, 54
Tomy Toy Co. 84, 87, 159
Tonka 84, 87
TOPIO (robot) 51
Topo (robot) 162, 163, 229
TOSY (robot) 51
Toyota Robot 52
Transformers (robot) 84, 86
transhumanism 9, 33, 34, 35, 40
transhumanist 32, 47, 48
transhumanist soldiers 48
transistors 139
Tributsch, Helmut 148
Trotsky 40
TRW 194
Tsukuba Science City, Japan 90, 93
Turk (automaton) 72, 74, 131
Twiki (robot) 222

Twilight Zone (TV series) 18, 20, 77
Two Panda Deli, The 174

U

UCAV 191, 192
Uchiyama, Shunichi 61
Underseas Explorer 177
Unimate (robot) 140, 178
Unimation 90, 140, 169, 206
United Nations 19
Universal Automation 140
University of California San Diego 66
University of Florida in Gainesville 148
University of Michigan at Ann Arbor 53
University of Rhode Island 91
University of Sheffield 194
University of Southern California School of Medicine 219
University of Texas at Arlington 91
University of Texas at Austin 90
University of Tokyo 224
University of Western Sydney 54
University of Wollongong 227
unmanned drone 191
Ursula (robot) 186, 187
useless eaters 25, 28
Utopia 45

V

Vaish, Dr. Diwakar 101
Vanderbilt University 147
Verbot 159
Verderame, Frank 191
Verne, Jules 105
Victory Enterprises Technology 208, 225
Viking Lander 199, 200
Villers, Phillip 174
Vinge, Vernor 21, 24, 116
Virtual Girl (book) 116
voice box 145
von Kempelen, Baron Wolfgang 72, 131

W

Wabot 91
waiter/waitress 174
Walking Head robot sculpture 54
Ware Tetralogy, The (book) 116
War With the Robots (book) 116
Waseda Robot 91
wasps, flies, other "insects" and birds (drones) 195
watchrobot 87
Watt, James 133
We Can Build You (book) 114
Weinbaum, Stanley 106
Weiner, Norbert 139
Weisel, Walter 209
Weizenbaum, Joseph 90, 220
Wells, H.G. 203
Westinghouse 169

What Sort of People Should There Be? (book) 51
Wheelbarrow (robot) 195
Whee Me (robot) 165
Why the Future Doesn't Need Us (article) 25, 45
Williamson, Jack 106, 112
Willie (mechanical man) 134
Wilson, Wayne 155
window washer 176
wind-up robot 83
Winkless, III, Nelson B. 205
Wired (magazine) 25, 26, 45, 56
Wiseman, Paul 58
With Folded Hands (book) 112
Wizard of Oz, The (book) 80
Wolkomir, Richard 222
Wonder Stories (magazine) 106
wooden cat (robot) 130
wooden otter (robot) 130
Woods Hole Oceanographic Institute 178
Woodson, Elizabeth 222
worker rights 23
World After Oil, The (book) 210
World Future Society 41
Wright-Patterson Air Force Base 91
Writing Child, The (automaton) 132

X

X-01 (robot) 128
Xbox Kinect 11

Y

Yamazaki Machinery 97
Young Frankenstein (movie) 120

Z

Zenchua Science and Technology Park 21
Zhengzhou 21
Zod (robot) 87

About the Author

Well-known author of three #1 national bestsellers, Texe Marrs has written 46 books for such major publishers as Simon & Schuster, John Wiley, Prentice Hall/Arco, McGraw-Hill, and Dow Jones-Irwin. His books have sold millions of copies.

Texe Marrs was assistant professor of aerospace studies, teaching American defense policy, strategic weapons systems, and related subjects at the University of Texas at Austin for five years. He has also taught international affairs, political science, and psychology for two other universities. A graduate summa cum laude from Park College, Kansas City, Missouri, he earned his Master's degree at North Carolina State University.

As a career USAF officer (now retired), he commanded communications-electronics and engineering units. He holds a number of military decorations including the Vietnam Service Medal and Presidential Unit Citation, and has served in Germany, Italy, and throughout Asia.

President of RiverCrest Publishing in Austin, Texas, Texe Marrs is a frequent guest on radio and TV talk shows throughout the U.S.A. and Canada. His monthly newsletter is distributed around the world, and he is heard globally on his popular, international shortwave and internet radio program, *Power of Prophecy*. His articles and research are published regularly on his exclusive websites: *powerofprophecy.com* and *conspiracyworld.com*.

For Our Newsletter

Texe Marrs offers a *free* sample copy of his newsletter revealing the secrets of the elite, unraveling globalist conspiracies, focusing on world events and exposing secret societies, cults, and the occult. If you would like to receive this newsletter, please write to:

> Power of Prophecy
> 1708 Patterson Road
> Austin, Texas 78733

> You may also e-mail your request to:
> *customerservice1@powerofprophecy.com*

For Our Website

Texe Marrs' newsletter is published free monthly on our websites. These websites have descriptions of all Texe Marrs' books, and are packed with interesting, insight-filled articles, videos, breaking news, and other information unraveling elitist secrets, delving into conspiracies and coverups, and exposing the elite agenda in opposition to American freedom. You also have the opportunity to order an exciting array of books, tapes, and videos through our online Catalog and Sales Stores. Visit our websites at:

> *www.powerofprophecy.com*
> *www.conspiracyworld.com*

Our Shortwave Radio Program

Texe Marrs' international radio program, *Power of Prophecy*, is broadcast weekly on shortwave radio throughout the United States and the world. Power of Prophecy can be heard on WWCR at 5.070 Saturdays at 7:00 p.m. Central Time. A repeat of the program is aired on Sunday nights at 9:00 p.m. Central Time. You may also listen to *Power of Prophecy* 24/7 on websites *powerofprophecy.com* and *conspiracyworld.com*.

Mysterious Monuments Enshroud the World With Magic and Seduction

A sinister and curious *Architectural Colossus* is exploding across every continent on earth. The United States of America is at the heart of this incredible surge. Monuments, statues, and buildings are being created everywhere with evil intent and magical purpose. Designed by visionary illuminist architects and based on knowledge of the occult wisdom, Masonic geometry and sacred numerology, this Architectural Colossus can be traced back to the antiquities of Mystery Babylon. Its secrets, once unlocked, point to an amazing and frightening future destiny for you, me, and the entire world.

Over 800 actual photographs and illustrations
Large format, a massive 624 pages · $40

"The builders of this strange architecture comprise an elite cryptocracy—a veritable who's who of the world's most rich and famous. The illuminist monuments built by these men are essential to the Grand Design. *Mysterious Monuments* documents and unmasks their Master Plan to seduce men's minds and catapult humanity into a New Order of the Ages."
— Texe Marrs

Order Now! Phone toll free 1-800-234-9673
Or send your check or money order (postpaid) to:
RiverCrest Publishing ~ 1708 Patterson Road ~ Austin, Texas 78733

Secret Signs, Mysterious Symbols, and Hidden Codes of the Illuminati

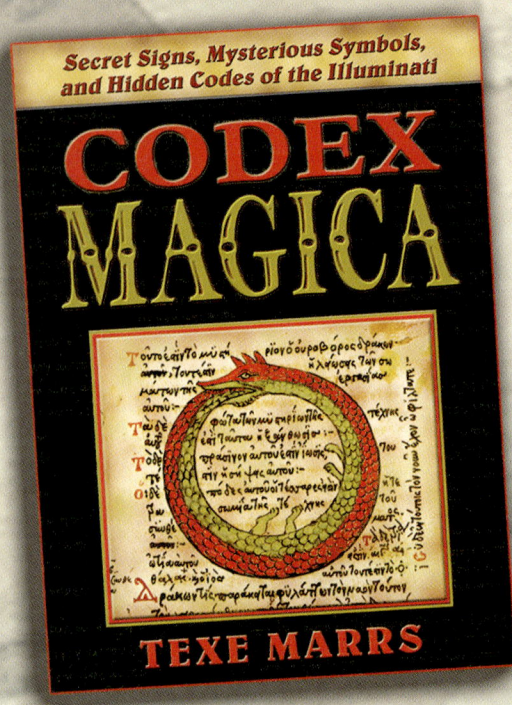

Codex Magica is awesome in its scope and revelations. It contains over 1,000 actual photographs and illustrations. You'll see with your own eyes the world's leading politicians and celebrities—including America's richest and most powerful—caught in the act as they perform occult magic. Now you can discover their innermost secrets. Once you understand their covert signals and coded picture messages, your world will never be the same. Destiny will be made manifest. You will know the truth and everything will become clear.

They Have Their Own Hidden Language

Over 600 incredible, mind-bending pages of never before revealed secrets by Texe Marrs ~ $40

Order Now! Phone toll free 1-800-234-9673

Or send your check or money order (postpaid) to:
RiverCrest Publishing ~ 1708 Patterson Road ~ Austin, Texas 78733